The Living Air

The Living Air

The Pleasures of Birds and Birdwatching

Aasheesh Pittie

Illustrations by Sangeetha Kadur

juggernaut

JUGGERNAUT BOOKS
C-I-128, First Floor, Sangam Vihar, Near Holi Chowk,
New Delhi 110080, India

First published by Juggernaut Books 2023

10 9 8 7 6 5 4 3 2 1

P-ISBN: 9788195996902
E-ISBN: 9788195996926

Typeset in Adobe Caslon Pro by R. Ajith Kumar, Noida

Printed at Manipal Technologies Limited, Manipal

For
Smriti

There was a child went forth every day,
And the first object he looked upon and received with wonder
or pity or love or dread, that object he became,
And that object became part of him for the day or a certain
part of the day … or for many years or stretching cycles of years.

'There was a child went forth' in *Leaves of Grass*
– WALT WHITMAN (1855)

For I have learned
To look on nature, not as in the hour
Of thoughtless youth; but hearing oftentimes
The still, sad music of humanity,
Nor harsh nor grating, though of ample power
To chasten and subdue. And I have felt
A presence that disturbs me with the joy
Of elevated thoughts; a sense sublime
Of something far more deeply interfused,
Whose dwelling is the light of setting suns,
And the round ocean and the living air,
And the blue sky, and in the mind of man;
A motion and a spirit, that impels
All thinking things, all objects of all thought,
And rolls through all things.

'Lines composed a few miles above Tintern Abbey, on
revisiting the banks of the Wye during a tour. July 13, 1798'
– WILLIAM WORDSWORTH (1798)

सर्वभूतेषु चात्मानं सर्वभूतानि चात्मनि ।
विज्ञाय निरहंकारो निर्ममस्त्वं सुखी भव ॥ ६ ॥

sarvabhūteṣu cātmānaṃ sarvabhūtāni cātmani |
vijñāya nirahaṃkāro nirmamastvaṃ sukhī bhava || 6 ||

Realizing the Self in all and all in the Self, free from egoism and free from the sense of 'mine', be happy.

— AṢṬĀVAKRA GĪTĀ 15:6
(translation by Swami Nityaswarupananda [1969])

Contents

Birding

Birders

My Kind of Birding

Foreword

I first met Aasheesh Pittie in a book, which was supremely fitting because the work itself was about the importance of bookish encounters. His *Birds in Books* (2010) is one of the great bibliographies of Asiatic ornithology, but it also celebrates how our real-world explorations of nature are continued and enhanced by our forays into the written word.

However, I got to know better both this bookish man and his writings while producing a work of my own. *Birds and People* (2014) is a global survey of the cultural meanings of birds, and it incorporates into the text the thoughts and reflections of more than 650 individuals from over eighty countries. I found that the most generous of all contributors were Indians and of those many kind correspondents, two were by a head and shoulders the most prolific. Aasheesh is one of this pair.

While we have never met and speak only occasionally, I've long known that he is an acutely observant man who is passionate about the living world. Both qualities shine through again and again in this lovely volume of essays, gathered from more than three decades of journalism and occasional pieces on wildlife in and around his home in Hyderabad.

They span an enormous range of themes and moods from the deadly seriousness of the Anthropocene epoch or the Sixth Extinction, to the hilarious way Spotted Owlets emerge blinking and bobbing from their roosts 'as though surprised that yesterday's landscape continues to exist undisturbed'. I also learnt a great deal. Sprinkled across the articles are prized nuggets of information: the fact that White-breasted Kingfishers eat young Little Grebes; that Little Grebes can squeeze the air from their fur-like plumage and sink their bodies below the water's surface; that Black Drongos will eat bats and even small birds.

Another key aspect of this naturalist is the sheer quality of his looking. He is the most patient of observers. Turn to the lovely piece on page 47 entitled 'Disappearing Dabchicks'. It describes his visit to a local pond, where Aasheesh became entranced by these gloriously ball-like waterbirds and before he knew it, the 'sun was balanced on the horizon … the disporting Dabchicks had engrossed me for three hours'.

Three hours spent looking at Little Grebes! How many of us could focus on a tiger for three hours without being a little distracted? The American ornithologist Ken Kaufmann, when asked once what made a great birder, said that he or she had to really, really love birds. By this measure, Aasheesh is one of the finest I know, but there is a wider point embedded in his capacity for concentration. He is as excited by things that are part of the everyday as he is by the rare and exotic. Here are essays on Jerdon's Courser and the Great Indian Bustard. Yet, they run alongside studies of the Common Iora and the Coppersmith Barbet. Because, truth be told, there are really no ordinary birds, nor ordinary life. It is all exceptional and Aasheesh Pittie is alive to this essential truth.

While he might write about many of these encounters with joy and loving attention, he is not afraid to call out the callousness and indifference which are at play in our systemic loss of the natural world. When he entitles an essay 'Can We Afford to Lose the Great Indian Bustard?' the question is entirely rhetorical. To the author, the disappearance of this magnificent species is nothing short of a national disgrace.

In another piece called 'The Death of a Nightjar', he describes how he discovers the corpse of one of these crepuscular moth-eating birds by the roadside by chance. Examining the beautiful, subtly cryptic plumage in typically exacting fashion, the author slowly realizes that there is a bullet wound on the body. To kill any bird involves an element of loss, but to kill one so harmless, in fact, one so helpful and amenable to humanity through its insect-eating habits, is nothing short of nihilism.

His book is full of good moral sense and wise judgement as well as fulsome descriptions of beauty and joy. Aasheesh Pittie is morally attuned to our times but also makes the perfect wildlife companion and I could not recommend his company – in books or in person – more highly.

Mark Cocker,

22 September 2022
Author and Naturalist

Introduction

I was introduced to and began birding in 1978, coming to it in a roundabout way, through reading. Posters and stickers were the rage those years, and the World Wildlife Fund-India (WWF-India) had printed large, endearing ones of a lion cub and a leopard cub. I bought these from my favourite bookshop, A.A. Husain & Co., run by a father-and-son duo in Hyderabad. Behind the posters was an address one could write to for membership. Through its quarterly newsletter, I learnt about Dr Sálim Ali and the Bombay Natural History Society (BNHS). Soon, I'd asked my father for a membership of that institution. Perhaps impressed by the imposing façade of the Society's 'Hornbill House', when he was next in Bombay, he obliged and also bought me a copy of Ali's *The Book of Indian Birds* (10th edition). I read about Zafar Futehally and his bimonthly *Newsletter for Birdwatchers* in that book, and wrote my first letter to him on 3 July 1979. His typewritten reply of 9 July came along with two old issues of the *Newsletter* – six of which would be mine for an annual subscription of ₹15. I was sold.

Birding fascinated me. By studying Ali's book, I learnt the

ropes of identifying birds. I exulted in the thrill of the outing, the chase so to say, and the satisfying entry of a bird's name in my field notebook – the veritable capture. Contributing to scientific ornithology was always a high point, though there were fewer opportunities then than there are today, what with the various special days when the world's birding family ventures afield to list and enumerate birds in their backyards, local patch, or nearby waterbody, and share these data on global platforms, in an unprecedented exuberance of Citizen Science. Despite its social nature, birding is essentially an individual pursuit. Like all birders, I too imbibe each bird in a uniquely personal way, saturating it with my own ethos.

During these four decades of birding, the activity has boomed in India, and it has also changed a great deal. Just as it did from the age of shooting and collecting specimens to that of watching birds through optical equipment without harming them, birding has now become part of the digital revolution and its incredibly active, worldwide social media.

Ornithology has benefited tremendously from the inputs of the globetrotting birder – mapping fresh distribution ranges, diving deep into the intricacies of systematics, even photographing a taxon in the wild for the first time. High-quality optical and digital photographic paraphernalia, available at affordable prices, has created an army of enthusiastic bird photographers all over the world. Travel has become easier, and birding tourism is on the uptick. Birders zip around the globe with shopping lists of must-see birds, using local guides and playback recorders to tick off species on their lists – another form of 'collecting'. Often, the primacy of a species' natural well-being plays second fiddle to the chimera of a top-class photograph or to the inducement of a response to a chattering call playback.

There are various degrees of transgression being committed through these acts that the perpetrators are blind to: consider the bird fleeing from a persistent photographer, or its energy sapping responses to artificial auditory stimuli. When we do not comprehend the impact of our seemingly benign actions upon our natural surroundings, we assume they have no consequences upon non-human life. I have often pondered the erosion of the birder's code, an ethic enforced increasingly in the breach, and wondered how birders could enjoy a finer, longer-lasting sensibility of experience.

Perhaps the trick lies in slowing down birding. In experiencing the bird in all its subtleties and behaviour by just being in its presence, in not forcing ourselves upon it. In allowing it to become invisible to us as it moves on, but remaining with it through the quiddity of its habitat, through its essence. Or perhaps it lies in reducing our desire to rapidly accumulate novelty, in relishing the slow-paced seasonal changes of a local patch. In accepting the non-human familiar as an integral part of a shared substratum and gently echoing the universal heartbeat of existence.

Over the years I have tried to make sense of such feelings by distilling the sentiency of my nature walks into 'words about birds'. Writing did not come easily to me. Page fright forces a reckoning not only of memory but also of the order of words that blacken the white sheet on the desk. It is a back-and-forth process – the more I write, the more I want to be in the field and vice versa. Gradually, memory speaks a language that plugs into the interconnectedness of things, justifying the process. Your way into this bewitching world may be through painting, or dance or even journaling.

The essays in this volume have been written over thirty-odd years, and some were originally published in diverse print and digital media. Some of the information they contain was pertinent at the time of their publication, though I have updated facts and figures where necessary. They reflect my joyous evolution as a birder, observer, thinker, and writer – a celebration of the non-human living world. They are arranged under four broad sections: the first contains my encounters with birds, the second is about birding experiences, the third has brief appreciations of three birding stalwarts who influenced me, and the fourth comprises ruminations on my kind of birding. A strict chronology, by publication date, has been deliberately avoided to allow a natural melding of articles written for varying audiences in daily newspapers, specialist birding journals, natural history magazines, online blogposts or in my personal journals.

Every birding excursion left an indelible impression upon me. The tangible result of those trips were the innumerable checklists I jotted in the field. But there was a much deeper engagement I had with my surroundings and whatever existed therein. Over time, these bird-centric outings developed a larger focus, a subtle recognition and embracing of a bird's Umwelt. On particularly prescient days, I became so absorbed in my surroundings that I seemed to merge with them when the buzz of the hyper-alertness that I had created by my sudden appearance melted away. I felt an overpowering sense of oneness with the world resonating deep inside me. My agitated mind, clogged with mundane civilizational preoccupations, quietened down. I felt like a creature amongst creatures, an existence amidst existence.

You may march to a different drummer on your personal journey. Whatever your pathway, I hope these essays enrich your birding experiences and inspire immersive expeditions into the natural world.

Birds

1

A Tryst with Jerdon's Courser

As the *Newsletter for Birdwatchers* approached its fortieth year of publication, Zafar Futehally, its revered founder-editor, wrote in his editorial, 'I suggest that some of you who can capture the right mood and the right memory, send me appropriately 300 words on what you consider has been one of your most exciting experiences in the world of birds. If we can procure a dozen of such contributions, it will make the November/December 1999 issue something of a treasure ...'[1]

Dr T.C. Jerdon, a surgeon with the East India Company, discovered a courser in 1848, which he sent to Edward Blyth, the curator of the Asiatic Society of Bengal. In his description of the bird, the latter stated that it 'inhabits the Eastern Ghats of the peninsula of India'.[2] Subsequently, the bird got its eponymous English name. Only a handful of specimens

9

were collected in the nineteenth century, the last sighting being from 1900! It remained a dubious member of the quartet, the Holy Grail, comprising 'extinct' birds of India, the dream of every birder for eighty-six years, till a small game shikari, Aitanna, of Reddipalle village near Siddavatam (erstwhile Andhra Pradesh), finally caught a bird on the night of 12 January 1986, and it was proclaimed rediscovered by Bharat Bhushan.[3] The euphoria that this discovery created in the ornithological world was temporal. A Jerdon's Courser has not been sighted since April 2008.[4]

On 21 February 1996, Rajeev Mathew, Mukund Kulkarni, Richard Grimmett, and I drove from Hyderabad to try and see Jerdon's Courser in Reddipalle, Cuddapah district, and the Great Indian Bustard in Rollapadu, Kurnool district. Rajeev, Kulkarni, and I had been birding together for nearly fifteen years. Richard had come to participate in an ornithological conference in India, and his monumental *Birds of the Indian Subcontinent*, co-authored with Carol and Tim Inskipp, was still under preparation – it was published in 1998, and the rest, as they say, is history. Miraculously, at 2220 h on that moonless night, we did manage to see a single Jerdon's Courser, and after birding its seemingly sparse landscape the next morning, drove onwards to the grasslands of Rollapadu, land of harrier roosts and booming bustards.

I sent the following piece.

Aitanna would hear none of it. He *had* to attend the wedding in the neighbouring village and the Kalivi Kodi[5] would simply have to wait another night.

A hired truck waited ominously, ready to carry the wedding party *and* Aitanna away from Reddipalle. After some more cajoling, he said that the batteries of his miner's lamp were discharged. But we had four-celled torches. He couldn't find the rattle that distracted the bird. We were willing to take that risk. As his resolve broke, I asked Richard to step out of the car and told Aitanna that he'd come all the way from England to see this bird. Would he have to return disappointed? Thankfully, Aitanna's self-esteem would not allow that to happen.

Aitanna led us into the pitch-black night, warning us not to switch on lights or talk. Our pavement-friendly feet had problems negotiating invisible stones and sudden depressions. But our hearts pounded with anticipation and our eyes strained after the dancing torch beam that Aitanna flicked haphazardly here and there. Two hours of this and we were trudging hopelessly. And then, it happened. The needle glinted in the haystack! A *Cursorius bitorquatus*, crouched on its long legs, stared at us with its abnormally large nocturnal eye. All our torches found their target. With bated breath we crept forward, afraid to blink lest the apparition vanish. A brilliant white supercilium separated its scalp from the face and neck and a double-lined necklace adorned its chest. We stood transfixed. As Richard reached for his sketchpad, Aitanna, the conscientious forest guard, motioned us away. It took us a while to get our breath back as we sat in silence, lost in thought. We had just seen the rarest bird in India, on one of the oldest geological real estates in the country, indeed even the world. As I realized that this terrestrial endemic had survived here for more years than our imagination allows us to register – but now faced an uncertain future – time stood still.

2

Can We Afford to Lose the Great Indian Bustard?

Considered rare even in 1947, the Great Indian Bustard's population has dwindled so alarmingly since then that it is now listed as 'Critically Endangered' in the International Union for Conservation of Nature's (IUCN) Red List of Threatened Species. Its current population hovers below 150 birds that cling precariously to their rapidly shrinking grassland habitat. Grasslands, often maligned as revenue wastelands, are victims of infrastructure development and agricultural expansion. An aggressive augmentation of irrigation facilities has changed their land use patterns from casually pastoral to protective agrarian, and the search for green or renewable energy, especially in the past decade, has seen a mushrooming of wind farms and their grid of overhead power lines in the vast, open landscapes of the bustard; power lines that the heavy bird cannot see during its cumbersome take-off, or while in flight, till it is too late for evasive action, resulting in a fatal crash. A conservation breeding programme was initiated in 2019, but it may be merely a fig leaf, covering too little and too late. As India blusters towards reintroducing the cheetah into its grasslands, one can only lament the travesty, the charade of prioritizing its grasslands for one species and denying it to another.

'Bashtad, saar!' Allahbaksh's downplayed guttural words brought me to an abrupt halt. Taking the telescope from him, I pointed it in the direction of his gaze and after a few anxious moments of squinting through the eyepiece, and twiddling the focusing ring, managed to crystallize a sharp image. An adult male Great Indian Bustard stood under a morinda tree. Stretching to his full height, he plucked ripe fruits from the lowest branches and swallowed them. A setting sun lit the bird in stark relief and through the scope I could see him very clearly. He walked around the tree as he fed. Once he stood for a couple of minutes behind the bole and peered around, as though aware of our presence, waiting for us to move on. When we stood our ground in silence, he must have realized that we meant no harm and commenced to feed again. Suddenly I realized there was an Indian fox scampering around playfully beyond the bustard! It spun like a top after its tail, breaking away to dart hither and thither in meaningless abandon. Fun and frolic are not the privilege of man alone. Momentarily distracted by the bustard, I looked away. When my eye rotated within the scope towards the fox, it had vanished. Meanwhile, the bustard, either sated or having consumed all fruit within reach, walked away leisurely towards the north-eastern boundary of the Rollapadu Great Indian Bustard Sanctuary.

His hangout was the 'humped' part of a flat-as-a-tabletop landscape and was a traditional displaying spot for male bustards. With his inflated gular pouch and erected neck feathers, the bird cocked his tail all the way over his back and began to strut in circles. I didn't see any females, but I was far away. He may have spied one or two on the other side of the hump and begun his act in silence. He did not utter the booming call that carries for half a kilometre. We watched him till the light began to fade. In that gloaming, a Red-headed Merlin began to hawk swallows, snatching them out of 'the living air'[6] with its delicate talons, and gorging on them atop a favourite granite pillar, which doubled as the sanctuary's boundary marker. In two days in the receding winter of the millennium's first year, we counted three bustards – one male and two females.

In the 1980s, when bustards were first spotted here and a sanctuary for them proclaimed, thirty-five to forty birds could be sighted within the span of a few hours. Rollapadu was considered the best bustard habitat on the Deccan Plateau, and on a good day, more than half a dozen males could be seen displaying from one vantage point! For a decade or so, the young sanctuary was protected, managed, and studied with zeal. Bustards were seen in good numbers. They bred, and their nests faced the natural vagaries of success and failure. Above all, individual numbers did not decrease.

The total protection of the habitat also boosted populations of other denizens. Blackbucks increased by leaps and bounds, and there are about 600 now. Indian wolves too increased, as must have other mammals like foxes and black-naped hares. In the resulting dynamic tussle for food and territory, subtle changes began to erode the ground from under the bustard's feet.

Species become extinct individual by individual, and this was the case here. There are many theories about the causes of this debacle – too many people and the resulting agricultural boom; too many blackbucks (as a result of total protection); widespread grazing within the sanctuary resulting in the increase of unpalatable vegetation for grasshoppers, the bustard's prey, and clandestine poaching being some. Also, a slackening of guard by those entrusted with the care of this irreplaceable natural heritage and national treasure – for it is not found anywhere else in the world. They seemed to have forgotten that protection is only the first step towards ensuring the survival of a threatened species, that the consequent second is active management.

This is the situation across the entire reducing range of the Great Indian Bustard. Hanging by the thread of human apathy and negligence is a member of a family that evolved 40 million–50 million years ago, perfecting a way of life on the short grass plains and arid regions of India. From its stronghold in the Thar (where its populations have halved since the 1980s), across Gujarat, Madhya Pradesh, Maharashtra, Andhra Pradesh, and Karnataka, the bustard is slowly but surely losing out. It has already been hounded out, by changing land use patterns, from Punjab, Haryana, Uttar Pradesh, Odisha, and Tamil Nadu. The bustard is not ready to die yet, but like Abhimanyu, it does not know how to escape from the chakravyūha (a military formation that encircles one) of habitat destruction that surrounds it. We are choking it to death.

The disappearance of any species from the face of the Earth is an irreversible tragedy whose finality cannot be emphasized enough. It's a physical loss for the natural world, a broken link in the proverbial chain of life. Increasingly, almost singularly these days, extinction is a direct result of human activity, the sixth of mass extinctions, the Anthropocene. By perpetrating these tragic and callous acts, we walk a one-way path that leads to the edge of a precipice from which there is no turning back.

Have you ever wondered what humanity is losing beyond the physical impoverishment that the loss of a species brings to an ecosystem?

To me, the Great Indian Bustard is as important a part of this planet as is a tiger or a tiger beetle, a Danaid eggfly, or an earthworm churning soil, breathing life into mud. The extinction of each species diminishes us and negates our collective intelligence. The bustard stands for the well-being of our grasslands and the myriads of life forms comprising that ecosystem. It strides through a landscape that gave character to my nation and to my brotherhood of Gujar, Maldhari, and Bishnoi.

In accepting its presence and rejoicing in its freedom, I can stand tall and be a part of the land that has sculpted the genesis of our agrarian lifestyle. The bustard struts in a wilderness that cannot be restricted to the insular and limited activities of human societies. It beckons the one human character that soars above them – our unassailable spirit. We cannot survive the twenty-second century successfully within the confines of our achievements alone. Our spirit yearns to walk with the bustard and to thrill at the jumping grasshoppers underfoot. To feel the wind's caress as it bends a sea of grass heads. To ruminate at the ebb of day under a gently burning, purpling, blackening sky. For deep down, we realize that the spirit of the land peoples our breath and ultimately strengthens our work. Its aroma is the only constant in our temporal world. It will not be denied.

3

The Silence of the Sparrows

When this piece was written, global sparrow decline was a topic of widespread concern and discussion. It would be pertinent to quote here what the 'State of India's Birds 2020: Range, Trends and Conservation Status' report has to say on this: 'Despite the widespread notion that the House Sparrow is declining in India, the analysis … suggests that the species has been fairly stable overall during the past 25+ years. Data from the six largest metro cities (Bengaluru, Chennai, Delhi, Hyderabad, Kolkata and Mumbai) do indicate a gradual decline in their abundance in urban centres. However, the extremely large range of the species across the country, and the lack of evidence for either long-term or current countrywide decline results in it being classified as of Low Conservation Concern.'[7]

My childhood was spent amongst sparrows. There were other winged commensals, too, that shared those golden days with me, but it was sparrows I felt closest to. Our house had an infinite number of crevices and cornices where these birds nested and a small garden and 'goshala' (cowshed) bordered with stretches of loose sand that they dusted in. Late afternoons were chosen for this surprisingly disciplined, communal affair. The fine sand had been warmed by a sun that slipped behind buildings on its journey to the other side of the world, leaving the dusting area in gentle shadow. Manik, the gaiwala (cowherd), would have finished chopping fodder for Lakshmi and Nutan, inadvertently scenting the afternoon stillness with the moist, resinous aromas of bruised verdure, setting those bovines into an impatience of nodding heads, flaring nostrils, and unbelievably widening eyes, till he poured it into their fodder boxes, barely avoiding spilling it all around as they nosed the flow of falling pieces. Not that it mattered, for their rough tongues later lapped up whatever dropped outside the box. It was between this time and an hour or so later, when he squatted next to Lakshmi, on his haunches with a bucket between his knees, to milk her, that the sparrows began their daily ritual, while the milk sang in the bucket.

Twenty to thirty cock and hen sparrows would gather on the ground between the henna hedge and the goshala, in an apparently meaningless knot, till a few of them shook off their listlessness by puffing up and ruffling their feathers, twisting this way and that on a patch of particularly soft sand, forming in the process a tiny rim of the fine powder around them. The sand billowed in minuscule clouds around them, passing through their feathers, suffocating irritating lice, mites, and other parasites. Dusting also absorbs excess oil from the feathers, which are then easily cleaned by preening, and restored to the flexibility required for aerodynamic flight. Their activity attracted others and soon there was a hierarchy of sparrows waiting their turn at the dusting bowl! Impatience and intemperance were put down with a sharp peck. Patience was rewarded by the duster hopping off, vacating the cup for another's ablutions. For the duster, it was a safe place in that moment of distraction, for others stayed alert for bored canines in search of fun, or a deadly Shikra, and shelter was close at hand in the spiky henna. On the days they chose not to dust themselves, for whatever unknown reasons, the remnants of the dusting patch, with its scattering of tiny depressions, remained on the ground, a microcosm of life's infinite rituals – meaningful to the observant, hidden from the preoccupied.

The Hyderabad of twenty–thirty years ago has changed from a city of minarets to one of massive billboards. In our hurry to transform the city into a modern metropolis, we forgot the house sparrow. Perhaps its ordinariness, its commonness, its dull omnipresence, like some old-fashioned traits – etiquette, sentience, an adab (respect, courtesy)

that was unique to Hyderabad – could not keep up with the blinding speed of development, the new culture of big, bright, and brash that seized us with a manic frenzy. Before we realized it, the sparrows had disappeared from our midst. Increasingly at parties, at street-corner gatherings, in offices, and at meetings, the hot question was, 'Where have all the sparrows gone? Why have they forsaken us?'

Those are not easy questions to answer. Sparrows have declined in many parts of India, as they have elsewhere in the world. They thrive in some cities – Mumbai, Delhi – but have almost disappeared from others (Bengaluru, Coimbatore, Hyderabad). In England, the population has dropped by more than 50 per cent since the 1970s. We Indians are not as meticulous about our data, and so we do not know the *scientific* trend of *our* sparrow populations. But we are aware of a reduction in numbers!

If this decline was a question of housing, of modern architecture lacking the extravagance of alcoves, cornices, crevices, and stucco versus older structures basking in their resplendence, there are the modern Indian cities that contradict the theory. If the presence of man, so essential for this commensal, was the question – for the sparrow thrives amidst the press of humanity – nobody can complain of a drop in human populations. What, then, is the matter with sparrows? Is it food, living space, or a combination of various situations? One can but conjecture and theorize.

Food could be a reason. Most of us now buy neatly packed grain from shops. Sometime ago, grain would be cleaned at home and the chaff scattered in the courtyard for birds. But sparrows do not exist by grain alone. They need insects and arthropods too in their diet, especially while raising a brood of ravenous nidicolous nestlings. The availability of such fare has been impacted by the use of pesticides in urban areas, private and municipal gardens, and within homes; their use has increased hundredfold in the past two decades. Their harmful effects are unrecorded scientifically but are graphic in reality.

Living space might be a reason – the lack of holes and cornices in modern architectural structures, our shift towards clean-lined, designer-book homes. How many of us tolerate a bird's untidy nest within our dwellings? How many let the web of a spider be? Who wants sandy places in these days of cement and concrete? The trails within most city gardens and larger parks are under threat of being cemented and paved! Or is it something inherent that was catalysed by our actions? Perhaps the sparrow needs the society of an unknown number of brethren to survive successfully. That could be a genetic prerequisite. If the depleting populations of certain species reach a critical low in their numbers, they are incapable of recovery. But what, finally, is the cause of this decline? There seems to be no single answer yet.

One thing, however, is certain. They have not forsaken us. We have them. The decline of the sparrow is a symptom of a uniquely human malady – our carelessness towards all other life forms save our own. This matters as we have the power to alter our environment like no other living force on the Earth does. And if we are careless towards life, can we take care of ourselves? Sometimes, I wonder which is worse – to be blind or to see and not comprehend?

4

The Death of a Nightjar

Returning from a tiring morning's birdwatching, we stopped under a roadside tree to freshen up. As I swung the vehicle off the road, I spotted a crumpled shape on the ground, amidst the dry leaves and stones, below the overhanging branches of a peepal. My first reaction was that it was a babbler until I went up to it and realized, with a sharp intake of breath, that the lifeless bundle of feathers was a nightjar. Knocked aside by a speeding vehicle, I thought as I crouched closer for a better look. The eyes that held night within their luminous vaults had been taken by ants. Its soft plumage, however, spoke of a relationship with an Earth we are rapidly losing sight of – the burnt browns and greys that intermingled seamlessly with brittle, fallen leaves scattered over dry grit-strewn ground; the rufous of deepening shadows reflecting a crepuscular firmament; spots of brilliant white mingling with the light that connects curving horizons and all life across this planet – horizons no longer visible to us in our myopic obsession with the immediate and the artificial.

Like most predictable humans, we too were interested in naming it – which was this species? We cannot seem to connect with our surroundings unless we reduce everything into logical, identifiable slots that fit our educated worldview. This is fine to a point, but we tend to stop with that clinical analysis, as though that were an end in itself, not looking beyond, at the interconnectedness of all life. We treat fellow Earthlings with foolish disdain, either ignoring their role in the existence of life on Earth or unable to comprehend its mysterious complexity. We walk the Earth with arrogance and yet are unable to control the results of our own actions.

We counted tail feathers, spreading them fanwise. Stretched wings. Enumerated primaries. Pushed open its pliant bill to reveal the cavernous gape and marvelled at the rim of stiff bug-snaring bristles around it. The inquisitive are oblivious to dignity whether in the living or the dead. Later, in trying to prepare a museum skin, we realized the cause of death. It had been shot; a neat hole punctured its lungs. But *why* would someone want to shoot a nightjar? It was not an item for the table nor was it good sport. But there are people out there who'll shoot at anything that moves, just for kicks.

A voice had been stilled, that once '*chuk-chuk-chukkurrr*'-ed in the gloaming, keeping the faith of an Earth that was winding down its operations for the day. A life snuffed out that did not interfere with our lifestyle yet became a victim of our senselessness. Had the nightjar lived, it would have flown up into the darkening sky, swallowing innumerable mosquitoes as it patrolled its territory, oblivious to the fact that it was 'helping' mankind in their battle against the mosquito. But animals have no foresight, nor do they mull over their actions. Unfortunately, neither do most of us.

5

ICRISAT Diaries

The International Crop Research Institute for the Semi-arid Tropics (ICRISAT) is a joint project in farming science, between the Government of India and the Food and Agriculture Organization (FAO) of the United Nations. It is spread over 1,390 ha, of which 800 ha is arable. A unique feature here is the presence of both red and black soils. It is a patchwork of varied habitats, with wetlands, fallow fields, cropland, scrub, and coppiced areas attracting a rich diversity of avian species. Here, birds are used to humans, and hence do not fly away easily, allowing for a Bharatpuresque ease of birding. However, birding is discouraged, and special permission is required for entry. As part of the Birdwatchers' Society of Andhra Pradesh (BSAP, and now Deccan Birders), I've been birding this appliqué landscape, isolated within a sprawl of industrial suburbs, since the 1980s, welcomed in the early decades by on-campus birders, Mary Peacock, followed by Dr Tom Hash, both keepers of the ICRISAT list. These diary entries originally appeared in my blog: Indian Courser.

Lark Song

An early morning sun haloed the landscape in golden light as I drove between fields of ploughed black soil towards the dump near Patancheru Cheruvu (a local pond), where I hoped to see a frisky Bluethroat prance after insects. Lark song suddenly poured in through the open window, and I pulled over to the side of the track. An Indian Skylark fluttered somewhere between the sun and me. I could not see it, but that did not matter, for it is one of those birds whose song surpasses its physical appearance in attractiveness.

I simply stood against the car, drenched in the glorious and profuse warbling emanating from his tiny, quivering syrinx. He drifted hither and thither on vibrating wings, exploding with the wound-up energy of his voluble performance. I wouldn't be surprised if his pinions fluttered with the kinetic fervour and excitement that consumed the little creature. I would like to believe that his ecstatic levitation and buoyancy were the result of that full-throated flood of uncontrollable sound rebounding from terra firma in aural waves and cushioning him in the ether.

Listening to him, all else faded away. I squinted into the sun, but the bird remained unseen, just his radiant melody flooded down, mesmerizing me with its repetitive strain, its slyly imitative descants and its clever improvisations. His stamina was monstrous. The performance went on and on, never reducing in volume, never slowing down, never faltering. Minutes passed and the aerial songster's luminous art abided. What a magnificent moment: to have stood still and listened to an invisible bird pouring his heart out through sunlit skies! To have spied his partner crouched beside an upturned sod, awash in that rhapsodic serenade! To have realized that the world's magic, its charm, its achingly simple joys are so easily within our reach. One simply must connect with nature or disconnect from artificiality.

When his time in the sun had run its course, his song ended abruptly, as though switched off, and he parachuted on cupped, outstretched wings, landing unobtrusively beside her. No one who didn't know better would believe that this superficially nondescript ball of feather was a virtuoso; that such a drab consumer of chitin and seed was so spectacularly endowed.

That is, precisely, the endearing charm of nature. It reveals its secrets at its own pace. Hurry it knows not, neither does it tolerate impatience. But silence, stealth, and solitude are handy at divining its mysteries.

[7 December 2014]

Magic and Mystery at the Marsh

As I eased on to the dirt road that is the ICRISAT campus's boundary with Patancheru Cheruvu, a Purple Heron stood transfixed midway upon a connecting track. Its posture was oddly comical; one leg in front of another, halted mid-step, the beak agape, its entire body stock-still. How odd was that! Normally, it should have taken off at my proximity.

Looking up the road I was to drive on, I saw three mongooses. The fur of one was all ruffled up, as though it had had a vigorous Turkish towel rubdown after a dipping. Two *Urva edwardsii* were on the right side of the track, and a third emerged under the chain-link fence that marked the boundary. The couple was intent upon something beyond the fence on Patancheru Cheruvu.

I followed their gaze and through the ipomea-veined links spied a female Marsh Harrier perched gingerly on semi-floating vegetation. She was clearly uncomfortable, shifting her position this way and that.

The mongooses looked distinctly unhappy, almost dying to complain about something. After hesitating for three or four minutes, they crossed the road with much raising and lowering of heads, taking furtive sniffs of the air, with minced, floppy footwork, strangely reminiscent of an indecisive hyena. Their attention evidently was riveted upon the harrier, which momentarily rose with an 8-inch limp fish hooked on its talons, landed on firmer ground, and commenced breakfast.

Minutes before this drama, a *Circus aeruginosus* was seen harrying waterbirds, flushing ducks and sandpipers. Clearly, it had been on the hunt awhile, as its aggression seemed to scatter the birds. But he had no luck with them.

So who killed the fish? Was it the mongoose trio, creating a ruckus over the prey that attracted the hungry bird that pirated away their prize? Or did the harrier come by the fish first and the mongooses simply coveted it? Did the heron at all play a part?

Every moment in the wilderness is filled with the drama of life. Luck, and chance, play their hands in revealing the poignancy of a pageant sometimes, which has its delights. Stealth, patience, and watchful silence are no less powerful tools in a naturalist's toolbox and provide the greater satisfaction to an inquiring mind.

The mystery of the behaviour of those four creatures remained. Perhaps with my intrusion into their landscape, I had perpetrated events that one or the other could not foresee and a third benefited from!

[7 December 2014]

Raptor, Interrupted

A Short-eared Owl-sized bird of prey flew away from me on the road bordering the eucalyptus plantation huddled at the extreme southern border of the ICRISAT campus. It must have been perched on one of the trees that lined it. Straining through the windshield for some distinguishing features on the rapidly disappearing rear profile, all I could register was an ochre brown wash concentrated in the tail region. Also, its secondaries were held perfectly horizontal, and the propulsion was from an energetic flicking of the primaries, carpal joint outwards.

I had stalled the car in my excitement at the thought of an *Asio* and could merely watch as it dipped into the leafy sanctuary about 100 metres ahead, my adrenalin draining as it disappeared into the trees.

Viewed upon the horizon from a distance, the trees were impressive, even grove-like. But up close, they had the frigid regimentation of a plantation. I wondered whether they stood as a windbreak or were prospective timber. There was no undergrowth. The trees stood like soldiers on parade, at branch length from each other. On the other side of the road, a meadow of wild thigh-high grass flourished.

I drove on, beyond the point where the bird had disappeared, switched off the car and climbed out into a partially tamed landscape. A railway line paralleled away not far beyond ICRISAT's boundary, and the combustion engine roared periodically on the outer ring road. But once I put that behind me, the land rolled away nicely without an object of artificiality, a scarce commodity in urbania and a solace for concrete-weary eyes!

A hen Montagu's Harrier coasted the eddies above the platinum-blonde grass with such effortless ease that one wondered whether that flamboyance was only the function of heart, muscle, and feather. Entranced, I watched her slide and turn and pirouette and albatross into the breeze to gain height and slip and curve to catch the invisible element repeatedly under her masterly sensitive pinions. All the time, her white-framed facial disc faced earthward, focusing owlish eyes and ears on the tiniest movement and breath that betrayed fear in the feathered or the furred. What a formidable combination of power and grace, of locomotion and concentration, of economy and extravagance!

I walked back, the way I'd driven, pausing now and then expectantly. But no owl flushed. I stood awhile, taking in the serenity, and then returned to the car. Just before opening its door, I looked along the ruler-straight treeline and from the apex of a eucalypt, midway down that line, a *Butastur teesa* blazed its white eyes at me. Motionless and ramrod straight, it stared me down from its elevated perch. This was the raptor I'd disturbed, and its displeasure lacerated me from that laser-sharp eye. Were those talons kneading its frustration into its perch? In a moment of heightened consciousness, in that silent pocket wilderness, I realized that despite all caution and stealth, my desire to mingle with the wild – shunned in mankind's quest to subjugate it – would always remain an intrusion in the world we walked away from.

[7 December 2014]

Air, Land, and Water

Barn Swallows peppered the skies. The mind fluttered at the thought that their fluid flight had power enough to propel those feather-light globetrotters across the crumpled and ruptured geography of continents, over the endlessly curving horizons of oceans, keeping faith with a genetic clock whose Earth-girdling pendulum swung them between the high points of procreation and perpetuity. And having arrived here, they stretched a horizontal seine, an animated mobile net over marsh, open water, and crops, snaring the massed aerial plankton in their flying maws!

In the black cotton soil stood three Indian Coursers. The field was prepared in neat rows, as most fields seemed to have been, and their accordion canvas revealed those smooth humped creatures on their bleached bone skeletal legs. The symmetry of their beings is such a powerful magnet for the aesthete birder. We glassed them, twisting uncomfortably in the car, whispering our awe to each other, scared our movements would startle them into flight. They too seemed frozen in fear or that supreme confidence that cryptic creatures have in their invisibility. The tension was palpable. Their tricoloured heads – black, white, and shades of sand – remained fixed at one angle for endless minutes. When they relaxed, it was the head they moved first, easing the crick in their necks; one cocked a black iris skyward, another looked away and I glimpsed the tri-coloured 'V' at its nape. Gradually, they began their comical dart–stoop–straighten–dart form of foraging, moving away from us imperceptibly.

Opposite the Red Tank (South), near the eastern perimeter of the ICRISAT campus, separated by the raised earthen

road, and at a much lower level than either, lay inundated paddies. They'd been freshly planted, at least two fields, and the sprouting crop of rice was still thin, with much soggy ground clearly visible.

Three Common Snipes stood in ankle-deep water like earthworks. When one tilted and inserted its straw length bill into the squelchy ground, did it sip up the earth and become dun-coloured? When still, they coalesced into their surroundings, gathering the mantling sky and the cradling earth into their protruding nocturnal eyes. In flight, bedazzled by daylight, they skimmed a zigzag course to escape.

Their hungry companions in the field were a few jittery and cautious Spotted Sandpipers. Both birds are more seasoned than the swallows as world tourists, returning to India once the monsoon quenches its bone-dry pelage and the depressions in its contoured landscape are softened by the liquid succour they hold. They formed the vanguard of the teeming flocks that would arrive as the cooler months progressed.

Treading a trembling floating world of plant and water, Purple Moorhens uttered shrill creaks, as though panicked by the sudden, unbidden sensation of sinking. They mimed a scuba diver crossing overland as they exaggeratedly lifted their reed-stalk toes high over recumbent soggy water plants, pantomiming stealth, and placed them cautiously upon floating fronds.

Several were on firm ground, twitching their stubby white tails up over their backs as they strolled in the thin shadows cast by reeds on the margins of banks. I saw one reach up to a seeding grass head with its beak, bend it to be clasped in the folded skeletal umbrella of its grotesquely long toes, and then clean the seeds into its mouth with a sideways swipe of its partially open crimson beak. Their feathers were a frenzied palette of the azure and verdure world, now dominated by dark hues of shadows, now bouncing the light of the sky through emeralds. Yet my eyes often missed their robust rotundity in their marginal world of land and water.

[12 October 2014]

6

Falcons in Focus

Driving towards the Red Tanks, a set of twin small ponds near the eastern perimeter of the ICRISAT campus, I glimpsed through the corner of my eye a grey arrow streaking across the landscape and disappearing into the crown of a toddy palm. CTH and I stopped on the bund, unfolding the spindly legs of our tripods, to scope a pair of avocets that had been reported from here around Diwali, in the first week of November. A small flock of waders huddled close to the small, fast evaporating pool. Ruffs stood about, some on the ground, sleeping with beaks under a wing, a few knee-deep in water, probing listlessly. No sign of avocets.

Swinging my binoculars towards the crown of the toddy palm, I spied the profile of a Red-headed Falcon. The tiercel would have seen our car long ago as he shot through the ether, abandoning a pouring landscape in his wake, to perch abruptly with a stone-hewn stillness in the palm. I could only see his bust over a frond. The hunter's large, all-seeing eyes, brown- and yellow-rimmed, reflecting the very world

37

it absorbed; the powerful curved beak; the chestnut hood and moustache. His white throat gleamed with bounced sunlight. He watched me then, in a casually alert way, boring his far more powerful eyes through the binoculars into mine. I could not hold his intense stare.

From his elevated, shadowed perch, he watched the flat landscape spread away all around, a quilt of undisturbed brown and ploughed red earth, of yellowing and green vegetation, of stagnant paddies and distant water. Small clusters of trees huddled here and there, and bare-branched thorny shrubs spoke an arid language. A bright sun shone from behind me in a wind-scrubbed, cloudless sky. The surrounding industrial hub was a noxious nuisance. If there were larks in the fields close by, they lay low, merging their browns with the furrowed earth. A Roller rasped in the background, not threatened by the hunter, but nervous in his presence.

Suddenly the tiercel rose, pumping his pointed wings with a surge of purpose and power. In the blink of an eye, he exploded from the sun upon the Common Swallow that had been zipping and unzipping the sky in pursuit of midges. At the last moment, the swallow tumbled out of the tiercel's scorched trajectory, glimpsing the brilliance of the sun in his blazing eye as it slipped past his flung anchor shape, miraculously avoiding the clutching talon. When I removed the binoculars from my eyes, pointing out the drama to CTH, we saw the falcon blurring towards her mate on closed wings. Astonishingly, the terrified swallow avoided the red-masked meteor as both raptors overshot the fluttering little feathered heartbeat. The element of surprise was blown, but the hunters pressed on. Again and again, the swallow escaped by a whisker, buffeted about in the frenzied violence of the attack, a panic-struck heart thudding frantically within its frail plumage. Predator and prey pouncing and prancing, locked in a tragic ballet of survival, spiralled to pinpoints in the domed firmament.

Fear, they say, gives power. Mercurial reflexes, honed in the pursuit of flying insects and by the sagacity of an Earth traveller migrating between distant shores, thrummed in the genes of the lucky swallow. In a moment, plummeting on curved wings, the falcon lit on the palm frond from where the tiercel had launched his attack, and he followed on her heels. The world swung back on its axis and surrounding birdcalls reached my ear again. CTH pointed towards a woodpecker. In that moment of distraction, the falcon slipped away. The tiercel continued to watch a warming landscape as swallows now hunted their winged prey between him and the sun.

7

Friends from Wingdom

Quite apart from the purely materialistic aspect, however, it must not be forgotten that man cannot live by bread alone. By the gorgeousness of their plumages and the loveliness of their forms, by the vivaciousness of their movements and the sweetness of their songs, birds typify Life and Beauty. They rank beyond a doubt among those important trifles that supplement bread in the sustenance of man and make living worthwhile.

> – *The Book of Indian Birds*, Dr Sálim Ali

The world is too much with us; ... Little we see in nature that is ours; we have given our hearts away, a sordid boon!

> – 'The World is Too Much with Us',
> William Wordsworth

Birds, it is said, can live without man, but man cannot survive without birds. Yet man, the super-ape, glowers from his pedestal, weighing the pros and cons of the continued existence of other life. When he passes judgement, it is generally myopic, drastic and suicidal.

All life, we know, is interconnected, forming an invisible chain, the strength of which lies in its weakest link. Every species is vital. But in our brawn (and the most developed of brains!) we break natural laws every day, everywhere. The irony is, we know a natural law exists, unlike a social law, only when we break one. There is a wide-ranging notion that other life forms threaten our existence, and we deal with such threats with characteristic ruthlessness.

Let us take just one branch of life from the myriads that exist – birds – and see how these glorified reptiles,[8] during their daily lives, inadvertently help us live ours better.

An Insectivorous Army

To catch the worm, an early bird must do a lot of work. Insects multiply at cosmic speeds, and they are found in every conceivable place – from ice–capped mountains to baking hot deserts, from mansions of the air to the expanses of the oceans, from dense tropical jungles to the warmth and safety of our kitchens. Their variety of form is astounding. Over 59,000 species are known to exist in India alone. Their menus include almost every form of earthly animal or plant, dead or alive. The statistics of their unchecked breeding are mindboggling. 'A pair of potato bugs, if undisturbed, would increase to 60 million in one season. And the infamous locust. Each locust lays several egg capsules containing about hundred eggs each. Once, 14 tons of eggs were collected from a 3,300-acre farm – estimated to produce 1,250 million locusts! Caterpillars consume twice their own weight in leaves every day."[9]

Man's fields are attacked, his home raided, his body assaulted, his very existence is threatened by these insects. We try to control insects with pesticides, herbicides and wormicides, but they soon become immune to chemical warfare. So, we must turn to biological control. Enter the bird. A major portion of its food is insects. Most birds raise their young on an exclusive diet of insects, many of which are harmful to man.

The House Sparrow visits its young nearly 250 times a day, its beak full of insects – and sparrows breed almost round the year. The swifts and swallows patrolling the skies constantly feed on aerial insects. The Roller – that flash of blue on telegraph posts – constantly wages war on ground insects which infest our fields. He is ably supported by the Drongos, which catch smaller fry (compared to the larger ones caught by Rollers) on the wing and occasionally on land.

Many birds also go directly to the root of the problem and destroy the eggs and pupae of insects. When a swarm of locusts settle down, a wide variety of birds, including mynas, drongos, bustards (the few that survive), White Storks, etc., feast on them. When you mow your lawn, you can see the Indian Robin probing for insects. Cattle Egrets lunge at insects disturbed in the grass by browsing cattle. In shady groves and forests, near torrential mountain rills, and in the darkest of nights there is always a bird, be it a thrush, flycatcher, dipper, or nightjar, continuing the crusade (from our point of view) against insects while going about its daily business of survival (its point of view).

Mao's Aide

Antagonism against any living thing, with only a superficial knowledge of its activities, gets one nowhere. Mao Zedong's aide once pointed out the large-scale destruction of grain by sparrows and the need to exterminate them. However, the massive surge in the insect population, which followed the sparrows' decline, caused more harm to the grain than the sparrows ever did. Swifts hunting near apiaries were shot to see whether they fed on 'useful' bees. Their gizzards contained only drones, which neither produced honey, nor had a sting![10]

The avian agenda often complements the human. Birds benefit us by destroying vermin. Rats, mice, and bandicoots live in fields and feed on standing crops; they multiply in grain godowns and finish off uncountable tons of valuable food. They make a nuisance of themselves in our homes and inhabit places that are cesspools of disease. Hyderabadi elders recall the years of plague with horrified shudders. But in the hierarchy of life, rodents find their predators in small hawks, owls, and of course snakes and smaller mammals. In our cities, Spotted Owlets and Barn Owls keep rodent populations in check – while in the forests and fields, there are other flying furies. The Indian Eagle Owl hunts three to four rats every night, descending on the creature as a silent messenger of doom. A single Sparrowhawk (a diurnal raptor), despite its name, hunts nearly 300 mice every year, and this makes up 80 per cent of its diet. A bird which kills even one rat a day suppresses the potential increase of 880 rats![11]

Scavengers

We all know the importance of sanitation. But our garbage cans overflow on busy thoroughfares. The least our municipalities can do is collect all the garbage and dump it in one place, far away from the city. From there, disposal is the work of scavengers. In the city, it is the crow and the kite; in the suburban areas, the Jungle Crow drops in; and beyond that, the vulture takes over. All of them are 'unsalaried public servants', And vultures are not 'different and dirty';[12] they have their own survival techniques. Unlike eagles, which squirt clear, vultures defecate on to their feet. Ornithologists believe their excreta is an effective antiseptic for their ordure and blood-sullied legs. The bald head that gets covered in gore exposes infectious bacteria to the purifying rays of the sun. They are so efficient as scavengers that they can dispatch a carcass within minutes. However, and tragically, the robust populations of vultures in India have become victim to a veterinary painkiller, and their population has plummeted.

Interdependence

With increasing levels of human population, deforestation seems to be the order of the day. Man does not care about the environment, but birds help man care indirectly. Birds are efficient seed dispersers – seeds pass through a bird's intestine, unharmed by the digestive process and sprout in favourable conditions. Most of the banyan and peepal trees one sees are naturally regenerated by such ingenious methods.

Birds are also efficient pollinating agents. Pollen gets dusted on to them when they probe flowers for nectar, getting widely dispersed in the process. Sunbirds have specifically evolved and adapted to feed on nectar. The coral tree, used for shade on tea and coffee plantations, is exclusively pollinated by birds. As Dr Sálim Ali wrote, 'A careful scrutiny would reveal that we are ultimately dependent on birds in this "House-that-Jack-built" sort of way for many more of our everyday requirements.'[13]

Man uses guano to fertilize his fields, he brings down their nests (swifts) to make soup, and he breeds them for consumption and sport. Birds also act as natural indicators of the health of our environment. A concentration of waterbirds promises good fishing, but that of coots might indicate choking eutrophication. In England, in years gone by, the canary was taken down into mines to test the safety of the air. Of late, however, indicators are rather dismal – a decline in raptor populations across the world indicates excessive pesticides (mainly dichlorodiphenyltrichloroethane, or

DDT) in the environment, which causes the shells of their eggs to weaken and break.

Ecosystem services that birds provide are so intricate and widespread, they cannot be easily comprehended, and so, the best philosophy is live and let live.

8

Disappearing Dabchicks

Some of my most delightful moments of birdwatching were spent one day in May 1984 near a small pond on the outskirts of Golconda Fort in Hyderabad, the erstwhile seat of the Qutub Shahi dynasty in the sixteenth and seventeenth centuries.

It was late afternoon, and the surface of the pond was touched with gold from a warming sun. I stopped beside this open sheet of water to observe the antics of some Dabchicks.

As soon as one of them saw me come a little closer, it disappeared. Just like that, *poof*! There was not a single ripple to dimple the water.

Soon, the same bird came up a little distance from where it had dived, alongside other Dabchicks! Then they settled down, and the fun began. One of the birds swam towards another, uttering a sharp, trilling titter. The other immediately turned tail and swam away rapidly. Then, the first bird rose above the water, beat its short wings wildly, frantically pattered on the surface and chased the other bird. Sometimes, there was a line of three or four birds, half swimming, half flying, chasing each other all over the surface of the pond. I finally roused myself when the sun was balanced on the horizon. The relative quiet away from the bustle of the city and the wildness of the disporting Dabchicks had totally engrossed me for three hours.

The diminutive Dabchick, or Little Grebe, belongs to the order of grebes known as Podicipediformes (the rump foots). Grebes originated nearly a hundred million years ago, and now the family comprises twenty-two species, divided among six genera.[14] Grebes are foot-propelled swimmers, with their legs placed way back in the body. This makes them awkward on land but eminently suited to swimming and diving. And are they masterly at that!

But unlike most web-footed waterbirds, grebes have stiff, horny flaps on each toe, known as lobate webbing. Grebes also differ from waterbirds in a few other ways. Their plumage is dense, rich, and lustrous. They have rudimentary tails, which do not facilitate steering in the air. But nature has its ingenious ways of combating such problems. Grebes use their dangling feet to manoeuvre in flight! Usually, these birds ride high on the water but are gifted with the ability to increase their specific gravity by expelling air from their bodies and from within their feathers, becoming less buoyant in the process. They leverage this gift by slowly sinking when suspicious. Sometimes, they swim along thus, with their heads and eyes periscoped above the surface, dubious and vigilant. This characteristic and mercuric diving abilities have earned the Dabchick its vernacular Hindi name of pandubi or dubdubi, meaning water diver.

The Dabchick is found throughout our country, in village ponds, water-filled burrow pits, lakes, rivers, jheels, or roadside pools. This 9-inch bird is seen bobbing on the water, rump feathers puffed out, looking rather blunt-ended and tail-less. Around Hyderabad, I have seen it on practically every stretch of water, with or without aquatic plant life, especially on the river Musi.

Dabchicks feed on aquatic insects, molluscs, fish, frogs, tadpoles, etc., which they capture mostly by diving underwater. Often, they chase an insect with their necks outstretched or hunt those lurking under aquatic vegetation. With the approach of April, most pandubis moult into their breeding dress and begin to form pairs. The head and upper parts turn dark brown, and the sides of the head a rich chestnut. The lower plumage is a lighter brown, and they also have a cream stripe at the base of the gape, which is visible from afar.

The pair formation ritual of grebes is generally quite

intricate, and that of the Great Crested Grebe is stupendously marvellous. A pair works like clockwork toys, diving, dancing on top of the water, exchanging gifts of vegetation, and the extent of their synchronicity is truly astounding. The birds share nesting duties, and while the parents are not particular about continuous incubation, they generally make it a point to cover the eggs with some loose material before they leave their nest. It has been found that the water temperature around such a nest is higher than in another part of the pond because the sodden vegetation gradually ferments, producing sufficient heat for incubation. After about twenty days, the precocious striped-and-spotted chicks hatch. They begin to swim immediately after their down has dried. But they dive amateurishly and can stay under water for a much shorter time than their parents. Sometimes, two broods are raised in a season. While the mother incubates the second clutch, the care of the first brood falls entirely on the male.

The plight of the young Dabchicks is pitiable when a White-breasted Kingfisher (their predator) pounces on them. Even if the adaptability of this bird to varying environments is good and its tenacity to life strong, how far will it tolerate our pushiness? Ecology is not restricted only to dense jungles; it is alive in our urban environment too. If 'in wildness is the preservation of the World',[15] then the habitats of local populations of pandubis will have to be saved. With increasing land acquisition, every jheel around the city is being reclaimed, every burrow pit filled up. Not only does this endanger our natural heritage but also our cultural and historical ones, for along with the flock of Dabchicks that frolicked under its shadow, the surrounding lands of Golconda Fort are once again under siege, this time by land sharks.

9

The Song of the Iora

The sound comes again, through a rash of yellow flowers on the babool tree, through its thorns and thin leaves. It is the song of the Common Iora. The tone is so pure and fluty! The triple notes of its descending whistle have a tranquil, melancholic ring, which stirs my heartstrings as I listen, spellbound, transfixed. It's a dreamy query which does not bring slumber. It rejuvenates the senses, rekindles the spirit, and sustains a vitality within me. It has a way of reaching the ear full, unblemished, unsullied, virginal. I feel as if a basic truth is being spoken, to no one in particular but for anyone who pays heed – it is there, in the open, on the babool, flaunted on the crest of the breeze. It is like so many truths of life which we take for granted just because they are there.

But these are the thoughts of men. This little black-and-yellow bundle of feathers does not know that he creates such upheavals of emotion within me. As he and his mate search the foliage for caterpillars and insects, they keep in touch with each other with an occasional *churrrr* and then the male lets out his sibilant, sliding whistle – *wheeeeeeee-chu-ooo* (the first syllable rising and the second and third falling), Sometimes, if in their rambles through the branchlets of a neem they are separated, the male drawls out a *wheee-too-too* (the descending whistle). Perhaps the female hears it as 'Here I am'.

If you are careful, you will see the pair, about the size of sparrows, moving through a tamarind tree in a garden, a mango grove on the outskirts of towns and villages, and frequently on a babool in secondary jungle. They are both quite similar in their non-breeding plumage – largely greenish-yellow. But the male always has a black tail, and can they carry those twin white wing bars with pride!

As the heat of the plains gradually diminishes and the skies lower their grey, brooding eyebrows, our male Iora undergoes a metamorphic change. A brilliant yellow replaces his greenish hue and a resplendent black cap, wings, and tail provide a fitting contrast. The white wing bars are now shining too!

He is ready to intimidate his rivals and woo his mate. This ageless ritual, much idealized by human emotion, is a response in our male Iora to various hormones in his body which goad him to do what he does. Scientists call this programmed response or, instinct, and the less scientific, the dance of love.

From a bush where the hen is perched, he springs forth into the blue, and with melodious, rich whistles, fluffing out his voluminous plumage, he descends in spirals to his mate's side, thus showing off his colours to their best.

When both are ready to settle down and raise a family, they build themselves an exquisite little cup of grass, abundantly plastered with cobwebs, fastening the whole structure on the upper side of a twig, 6 to 12 feet in the air. In this nest, the hen lays two to four pale, pinkish-white eggs, blotched with a purplish-brown. Both birds share parental duties when the eggs hatch. This state of affairs extends chiefly up to September, but there always are local variations.

So, when you next hear whistling notes outside your window, do not dismiss them as some idle schoolboy. It is perhaps the song of the Iora, trying to remind us that in the world he too has a place.

<h1 style="text-align:center">10</h1>

The Kotwal of the Countryside

You will see him often in the Indian countryside, though he frequents city gardens too. He is the glossy, black bird with the forked tail and a white spot in the corner of his beak. We call him the Black Drongo or by one of his evocative native names, depending on which part of this ancient land we dwell in. For the drongo peoples the length and breadth of the Indian subcontinent in three races. Hyderabadis know him as Zulfikar because his tail is shaped like the sword of the Prophet and perhaps also in recognition of the protection he provides 'weaker' birds when he is nesting! North Indians call him kotwal, for he is the farmer's knight in black shining armour. And since the drongo is so much more well mannered than a crow, he has been honoured with the title of King Crow.

To a layperson the drongo seems to be an evolutionary mix of a shrike and a flycatcher to occupy its unique niche. He has all the ruthless power of a carnivorous bird and executes this brawn with the finesse of a pirouetting flycatcher. Observe him chasing an invisible insect. With what delicate fluttering he pursues it, tip-turning as the little one side slips, coming up neatly from below and snapping his beak shut! Sometimes, the morsel is a trifle larger – so we think – but our acrobat has many a trick up his wing. He will bring his catch to a perch and batter it left and right until the thing is quite dead and broken – only then will he gulp it down. He enjoys this game so much, his exuberance is so unflagging, that you will see him at it throughout the day. He is among the earliest risers, astir well before dawn and only going to roost after dusk.

You can see him outlined against the sinking, swollen orb, a black speck on a fiery red canvas, rising occasionally, after a moth or other crepuscular insect and then plummeting back to his perch with closed wings and a whoop of victory. Having made the countryside his parlour, he has become an invaluable asset to farmers. With a palate for grasshoppers, locusts, beetles, bugs, ants, termites, and bees, he faces no complaints from them. When food is aplenty, a congregation of drongos may be seen feeding in conviviality. Sometimes, the drongo relishes a more sumptuous mouthful and preys on lizards, small birds, and bats. Yet, he is not satisfied. When the trees bloom, he visits many a tender flower and partakes of their nectar.

You will often see flocks of drongos on resplendent silk cotton and flame of the forest trees. Undoubtedly, the bird also helps pollinate a number of these plants. You should not be surprised when I tell you that our friend also indulges in unabashed piracy now and then. He chases foraging birds with great determination and forces them to release their prey, whereupon he grabs it in mid-air and carries it off. He uses various animals and birds, like bovines and the Great Indian Bustard, as mobile perches to give him a great view of the surroundings and the opportunity to snap up any insect his 'vehicle' frightens into flight!

The usual call of a Black Drongo is a defiant *ti-tiu* whistle. But beware of his mimicry. He comes very close to the mastery of his flamboyant cousin, the Greater Racket-tailed Drongo, in this art. The calls of commoner birds are uttered with the ease of an expert, but sometimes he will venture into the cacophonic world of man-made sound. I once heard him imitate a lawnmower as a mali cropped the lawn!

Like so many insectivorous birds, the drongo too nests in the monsoon when insect life peaks. Though the breeding season stretches from April to August, most nesting activity takes place between May and June. During this period, two or three birds will sit close together on a branch and 'talk at' each other in loud, harsh notes, bobbing the front parts of their bodies and raising and lowering their beaks emphatically. A bold bird, he builds his nest in a fork of the outer branches, in trees like tamarind, mango, babool, sheesham, etc. Generally, the nest is 4 to 12 feet above the ground and the nest-tree usually stands alone so that he has a good view of his surroundings.

We Hyderabadis see another cousin of the Black Drongo in the winter, the White-bellied Drongo. He is a seasonal, local migrant, who prefers shady spaces and the edges of clearings, but he'll come into gardens too. He resembles the King Crow except for his debonair white belly.

When next you go into the countryside, or pass a quiet, secluded pond between a grove of trees and paddy fields, pause for a moment. You might see a Black Drongo dive into the water to bathe and rise with a flutter of wings and scatter of diamond water. He will be silhouetted against the light and the drops of water will sparkle in the rising sun, and your day will be made!

11

Master of the Urban Skies

Who among us has not been fascinated by the aerial antics of a Black Kite? I have often stopped to marvel at this lithe hawk plunging through a maze of overhead wires into narrow lanes lined by houses and crawling with traffic to pick up, faultlessly, a dead rat from the garbage dump. And any schoolboy will tell you how he was left bewildered at lunchtime by the sandwich in his hand having disappeared in a whoosh of air. Again – twisting, turning, successful – the Black Kite!

Having evolved as a scavenger, the Black Kite finds enough food around human habitations to satiate himself, thus becoming a commensal of man. His long, thin, pointed wings and forked tail carry him effortlessly over houses; they help him dip down to roads, glide around fields, and sail over tanks as he searches for food, constantly turning his head from side to side, scanning every inch of the ground. One point of interest here is the tail of this bird. In our skies, only the Black Kite has a forked tail amongst the birds of prey of this size. The tail is used incessantly as the kite wheels and tip turns in the sky with the manoeuvrability and grace of an ace flier.

What sets the cheel (the onomatopoeic Urdu word for kite) apart from the other two common scavenging birds of our towns and villages – the vulture and the crow – is his habit of 'lifting' the titbit and carrying it away to a private place where it is promptly dispatched. While members of the vulture family have a penchant for squatting on a dead cow and the crow for straddling a dead bandicoot in the centre of a road, the Pariah Kite carries away the morsel to favourite feeding perches. One often sees a kite being chased by a murder of House Crows when it has some food trailing from its claws. But very seldom are the crows successful in their 'piracy'! The cheel sometimes transfers smaller food from its talons to its beak mid-flight. Though the diet of a kite is not as catholic as a crow's, it is quite varied. Scores of these birds are seen near garbage dumps, slaughterhouses, bazaars, and dockyards, where they pick up offal with blistering speed and agility from under the beaks of the local corvid baddies. When breeding, a pair might turn to the poultry yard and become chicken lifters. Sometimes, they can also be seen sauntering on rain-dampened ground or lawns on the look-out for snails.

Often, in the evening, kites soar in large numbers. The purpose of this behaviour is not known – unless they do it for pleasure! Indeed, if you watch them circling, you will notice one break off and tumble upon another with partly closed wings, while the latter turns turtle to fend off the onslaught with his talons. No harm is done to either and they seem to enjoy it immensely. As each day ends and dusk falls, the kites wing their way in a particular direction to roost. These roosts are usually in large trees, where scores of birds gather for the night and there is much flapping and bickering as darkness envelops them.

As the cooler months approach, the cheel begins the chore of raising chicks. Its breeding season is quite long, lasting from September to April. During courtship, one bird speedily chases the other and dives at it constantly. The latter parries the 'attack' by turning over in mid-air and extending its talons. Sometimes, they interlock talons and tumble from the sky, screaming loudly, only to separate within inches of rooftops. Their loud, musical, ringing calls – *ewe-wir-r-r-r* – are heard almost through the day at such times.

Their nest is built in the branches of large trees, 7 to 14 metres above the ground. Both sexes collect sticks, wire, rags, plastic, and all sorts of rubbish and pile them up into an untidy platform. Two to three eggs are laid and when the female (generally the larger bird) incubates them, her partner brings her an occasional rat. He stands guard over the nest while she gorges herself. When the eggs hatch, both parents help feed the young. The chicks fledge in about forty-five days but continue roosting in the home tree for another three weeks. They differ from adults in being more spotted, and often their tails are not as neatly forked as their parents' – nature's little way of confusing new birders.

With the decline in the vulture population since the late 1990s, a portion of their niche, feeding by scavenging, has been usurped by kites. Garbage sites are teeming with them. However, we must remember that they are at the top of their food chain and will suffer the consequences of what they ingest. I say this because almost everything they consume, whether a by-product of animal husbandry (and the drugs we feed these animals to bolster human nutrition) or something that naturally occurs in the environment, would have been the victim of the toxins that we pump into the environment, which then accumulate in the birds' tissues. No one knows how that will eventually affect kite populations.

12

Vulture Culture – Not All That Offal

In the late 1990s, scientists discovered that the vulture population in India was in serious decline. The non-steroidal veterinary drug, diclofenac, used as a painkiller, was identified as the tragic nemesis of India's Gyps vultures, gutting their population by 95–99 per cent across the land, catapulting their status to IUCN's Critically Endangered category. The socio-economic consequences of this cataclysmic descent towards extinction are unpleasant to say the least. The vulture's niche has been usurped by feral dogs (and to a small extent by Black Kites and corvids), whose population has burgeoned throughout India. They have no predator to control their numbers, and they often fear none as they form formidable packs when aggressive or threatened. Lacking the powerful digestive systems of vultures, feral dogs are susceptible to the diseases that rotting carcasses harbour, becoming rabid and dangerous to humans, suffering painful deaths without recourse to veterinary care. An elaborate captive vulture breeding programme across the country aims at creating a pool of birds that will be released into the wild once the threat from diclofenac subsides, and this has begun in a small way.

This essay was written in 1988, when, happily, our only concern was how we perceived vultures, given their unsavoury habits.

We can't afford to look down upon the vulture, for it is certainly more efficient than a municipal squad! While the ungainly bird may never win a beauty contest, it does a quick job of clearing carrion and preventing disease.

In fact, the vulture is more sinned against than sinning. Projected as a dweller of barren and deserted realms – where life seems overwhelmingly mortal, alone, helpless – it fills the role of the lurking predator, patiently awaiting the inexorable end of the hapless. Being associated with the mysterious and inexplicable, it has been accepted as the harbinger of death and suffering by a society that does not want to inquire into the whys of the so-called lesser forms of life. Let us, for one minute, cease to emulate the ostrich and evaluate the vulture and its place in the modern world.

The vulture is a scavenger par excellence. Can you imagine an Indian countryside where the dead, be they human or animal, are left on the outskirts of villages or along riverbanks without these natural municipal squads? Indeed, a good deal of the work of urban municipalities is also done by vultures.

E.H. Aitken, writing more than a hundred years ago, had said, 'Of all the unsalaried public servants who have identified themselves with this city [Mumbai] and devoted their energies to its welfare, no other can take a place beside the vultures. Unfortunately, the vulture has never lent itself to the spirit of heraldry.'[16] In the wild, vultures scavenge upon dead animals, hastening biological processes by converting essential nutrients into energy with great speed. That they have survived millions of years of evolution is proof enough of the usefulness of their methods. They are the distilled quintessence of aeons of natural selection.

A day in the life of a vulture may begin with a sunbath. Early one morning, at the municipal dump near Mahavir Harina Vanasthali, Hyderabad, I saw about fifty White-backed Vultures on the ground. They all faced westwards, leisurely sunning their backs. It is said that the soaring flight of these large birds bends their feathers to the extent that flight becomes uncomfortable. The sunlight straightens their feathers in four to five minutes, a job that would take two to three hours in the shade!

When the sun is fairly high up in the sky and the earth heated enough to send up spiralled draughts of hot air known as thermals, vultures take to the wing. Being big and heavy birds, they require a tremendous amount of energy to fly by continuously flapping their wings. Hence, they make use of these thermals to soar. Around mid-morning, on the outskirts of towns, one often sees vultures soaring on stationary wings, round and round inside a thermal current. When they gain sufficient altitude, they wheel towards another thermal and repeat the process. This way of travelling enables them to cover great distances daily, and they utilize as little as one-third of the energy required for normal flight!

The vulture searches for food while it coasts through the azure expanses of the sky. There are some vultures that spot food indirectly. White-backed and other vultures of the Old World watch each other and the behaviour of smaller avian scavengers while flying. If each vulture flies at least a mile or two apart, keeping an eye on the others all the while, no sooner does one spot something suspicious in the behaviour of crows or kites or even treepies lower down in the sky – or for that matter, spot a carcass itself – and closes those ponderous wings to plummet towards the earth to investigate, its vigilant neighbour breaks off from its effortless, gravity-defying vigil and gives inquisitive pursuit instantly. This behaviour-sparked message is transmitted across miles of sky, and within seconds, before you can hear the Blue Bottle's buzz, the vultures are dropping from the sky faster than you can count!

Some vultures of the New World have extremely acute olfactory sense and can locate putrefying offal and carrion by following their noses! Enterprising and enlightened engineers in gas companies in America pump pungent gases, like ethyl mercaptan, through natural gas pipelines and then drive along them. Wherever they see a congregation of wild Turkey Vultures, they know the pipe has a leak!

As is prevalent in other social orders, vulture society also has its own hierarchy or pecking order. Sometimes, we see a gathering of three to four species of vultures gathering around a carcass. They may vary in size or structure. Though four species have collected, there may be just one that is adept at opening a whole carcass. Others have unfeathered heads and necks to enable them to shove deeper into carcasses through a small rent in the animal's skin as they gormandize! There is always one bird which comes in just for the bones that remain after the orgy! This last, the majestic Lammergeier – the bearded monarch of the immense solidity of mountain ranges – has evolved to crack bones and extract the marrow by taking them to a height in its feet or bill and dropping them on to a flat rock where they splinter.

The Egyptian Vulture is another smart bird. With a relatively weaker beak than its cousins', at times it must find alternative sources of nutrition. During the summer, in the Keoladeo National Park in Rajasthan, it has been observed walking up to and flipping over turtles trying to escape drying ponds. It then inserts its thin beak between the testudine's armoury, enjoying a quick meal. In Africa, these birds have been known to break open ostrich eggs, which are more than a mouthful, unlike other smaller eggs which they hold in their beaks and throw down to crack open. To crack the larger ostrich eggs, they pick up a stone, and holding their heads high, drop it on to the egg till it splits!

Is the vulture really a dirty bird with loathsome habits? Consider this instead. The bird's bald head and neck and bare shanks which when smeared with gore can easily be washed and dried in the sunlight, destroying all harmful bacterium. The tarsi are also disinfected when the vulture defecates a slurry white liquid comprising uric acid on to them. Its faeces serve the dual purpose of keeping the bird cool as it evaporates – just the way sweat works for us! Some carcasses have decayed so much before a vulture finds them that they are awash with enough bacteria to kill most animals. Vultures have heavy-duty detoxifying digestive systems which convert such putrid matter into energy!

At a carcass, vultures may not get a leisurely meal. There are too many of them, and there is much jostling and pushing and many an imprecation exchanged between birds. Each bird has only one stratagem in mind – it grabs as much as it can before it is dislodged from its place beside the carcass by another! Sometimes, there are only a few birds at a carcass, and they engorge themselves so much that it becomes almost impossible for them to take off!

Since meals are unpredictable, vultures must conserve as much energy as they can. It is believed that at night their body temperature drops a few degrees – another reason for that early morning sunbath! Their style of flight also conserves 70 per cent of the energy that would be required in normal flight. Between meals and on unfavourable, overcast days, vultures counter the resulting energy crisis by simply doing nothing. They cluster on their favourite roosting trees, perched hunched and forbidding, banking energy.

Of late, vultures have been attacked by the aviation world. Many an expensive aircraft has been grounded for repairs, some have even crashed, when a low-flying vulture is sucked into the engines. But this problem has been caused by humans. We locate airports on the outskirts of cities where tanneries and municipal dumps (invariably) exist, which are both frequented by vultures. All the problem needs is coordination between two or three government departments. Either tanneries and garbage dumps should be run on more scientific lines, or they should be moved elsewhere.

Vultures deserve more than we allow them. Cast in the forge of nature to occupy a niche that has been taboo to humankind since its emergence as the supreme animal, vultures are doing a great job. Can we afford to look down upon this bird, which, by consuming offal and decaying matter (thereby preventing dangerous diseases), does a great service to our society? Let us stop complaining that the vulture has no culture, that its table manners are offal! Let us try and begin to understand the way nature works instead.

13

Owl Therapy

The sun was lowering on a day when the chips were down. I sat at my desk, head in my hands, nursing a dejection bordering on depression. The blaze of orange on the panes of my windows – generally a balm for frayed nerves – seemed to radiate the agony of a sinking sun. A crazy pandemonium within my cranium threatened sanity. I seemed to be touching the margins of my existence.

Lifting my head to look out of the window, I realized that a shrieking discord was radiating upwards from our compound. How strange! My despondency had capillarized into the surroundings. A cacophony of terrified and angry birds drew me to the open window, and a large, dark, owl shape clutched at my being and led me down into the garden. I followed the sounds of avian disapproval and alarm from tree to tree till they coalesced into the darkening shadows of a giant peepal standing in the northeastern corner of the grounds.

Within the fragile safety of a million still leaves (the

humidity was 75 percent) perched a Mottled Wood Owl. Five species of birds were screaming 'ware-hawk for all the world to hear, forcing this shadow of the night to seek sanctuary. The House Crows were trying to pull its tail feathers in hesitant, jittery sallies. The White-headed Babblers, fluttering their open wings, hopped on branches all around, uttering their panicky, shrill alarm calls. A Black Drongo buzzed the owl-smudge frequently, careful to keep a margin of safety for evasive action. The Common Mynas grated and shrieked their hatred with forceful venom. And the treepies! They actually dove from behind and hit that feathered edifice repeatedly on the head!

I stood in a feathery mizzle, with mosquitoes pinging around my ankles, in the middle of a vegetable garden, mesmerized. Mine were not the only problems in the world! Here was a relic whose body held the essence of patience, tolerance, and wisdom distilled over aeons far greater that the span of my temporal troubles. Two dark, serendipitous orbs gazed down at me, riveting in their serenity – with the stillness of deep forest pools. Was this a lesson in flesh and blood? Alone, the owl bore a humiliating tirade from fellow aves, hunkered into itself, consumed by the flame of survival burning within it, assured by an instinctive wisdom that this too would pass! As light merged with darkness all around, I realized how true this was. By and by the oppressors went away, leaving the 'wise one' to its ways.

There was another thing that caught my attention. Many Black Kites roost in the big trees on our grounds, and they were disturbed either by the arrival of the owl or by the racket created by the smaller birds – I am not sure which. But all of them were seen 'floating' low over the peepal in which the owl suffered its tormentors. The kites flew in tight circles over this tree as they had done over the couple of trees in which the owl had sought refuge before alighting upon the peepal branch. Silent all the time, they left after a while, realizing that the owl had decided to stay put. These two 'signals', I realized, could be used by birdwatchers to locate owls in urban areas. My step was light when I started up the stairs and a brightly lit room welcomed me. The word 'therapy' was added to the many advantages of a beloved hobby that day.

14

Watching Shikras

Summer is heralded by lengthening days and leaf fall. It is also proclaimed by the strident bisyllabic mating call of the Shikra. These little hawks are found in every urban or rural neighbourhood of our country. The male is grey on top with black wing tips. He has yellow or orange eyes and yellow legs and feet. His ventral region is cream, closely barred with rust. Down the centre of his throat runs a thin grey stripe. His lady is larger than him, as is the case with most birds of prey, and brownish-grey on top. Their young are more like her but have broad brown vertical streaks and spots on their underparts.

Courtship is a time when the birds call a lot. Their usual *tittiyu* changes slightly to *tiyu-tit* during this time. They are most vociferous in the mornings. The male performs a courtship flight by deliberately flapping his wings in slow motion, raising his tail slightly to display his white rump. Together, they soar in circles, sometimes breaking apart and making two separate circles, only to rejoin and soar as a pair once again. They show off by diving steeply towards a tree, coming to a dead stop on a branch, literally standing on their tails! Then, they let out their piercing breeding call. Occasionally, they will tumble out of the sky like pigeons and disappear into a leafy tree.

Their high-rise penthouse nest, constructed by both birds, usually stands 5 to 17 feet above the ground in the fork of a stout tree or at the base of a palmyra palm's leaf stalks. They tug at and break dead twigs and small branches of trees that are then transferred to the chosen fork in their talons. When the female incubates , the male brings her food.

The Shikra is a generalist and feeds on a wide variety of animals. Surprise is its weapon – ensconced within the recesses of a leafy tree, it keeps a sharp vigil on the movements of possible prey below. The attack is fast and fatal – the bird pounces upon its unsuspecting victim, killing it with its talons. Often, it will harry a flock of small birds trying to hide in dense bushes, invariably picking up a weak laggard. At other times, it will cause squirrels to jump out of their striped skins as it chases them round and round the trunk of a tree. The petrified fall prey and the quick escape! The bird is known to pluck a Pipistrelle from the air at dusk, when the little echo-locaters emerge from the crevices in buildings or the folds of palm leaves. Garden lizards are one of its favourite savouries.

When the Shikra zips through a maze of tree branches or crosses from one grove of trees to another, interspersing the flapping of its wings with short glides, little birds and animals shout out to each other, *'ware-hawk* as though Gabbar Singh was on the prowl! The enjoyment of this bird does not require binoculars. You just need curiosity and the ability to derive pleasure from nature. Keep an ear open for the Shikra's call, and you'll hear it in any locality. Rush to a window or go outside if you are within walls, or just look around if you are outdoors and allow your ears to show you the bird streaking through the canopy, spiralling on thermal currents in the sky, or perched among the shadows of a tree's crown.

15

The Spotted Owlet: The City Gossip

As life slows during those magical moments we call twilight, and a brilliant burst of pink, purple, and blue merges into the inky black fabric of night, the family of Spotted Owlets that lives in my compound shake awake from their diurnal slumber. They have slept all day in a hollow inside the trunk of a dead coconut palm, and they now appear at the entrance to begin their 'active' life. They come up, one at a time, pausing to view the world through large round yellow eyes. Lazily, they flap to a branch, where all four huddle together, stretching their wings, twisting their heads around to breaking point, blinking their eyes with dignity, bobbing on their short legs, as though surprised that yesterday's landscape continues to exist undisturbed!

In the deepening dusk, they fly to a lamp

post, street light, or overhead wire, uttering their screechy chatters, *cheewak-cheewak-cheewak*, bobbing their heads to zero in on different objects with their binocular vision. Their call, carrying clearly over the roar of vehicles, barking dogs, and other neighbourhood sounds, is the best way to identify them. A variation is a harsh *chirrur-chirrur-chirrur*. Once familiar with these calls, you'll be pleasantly surprised to find the bird's presence in the most unusual places. Many years ago, while listening to a concert by Pandit Ravi Shankar and Ustad Allah Rakha in the football field of Hyderabad's Lal Bahadur Stadium, I heard the call of an owlet above the strains of the sitar. Sure enough, a bird perched on the top of a wooden post near the shamiana (a temporary tent), perhaps attracted by Panditji's magic! The odds, though, would have been that it was taken aback by the strange disruption of its usual hunting ground!

The Spotted Owlet can be heard and seen all over Hyderabad. Whether you reside in the busiest areas – Begum Bazaar, Sultan Bazaar, Charminar – or the more open localities – Jubilee Hills, Marredpally, etc. – this little owlet is found everywhere. It flies over large areas in the night, hunting its favourite prey of large insects, moths, and mice. With the increase in human populations, pests like rats and mice have also increased hundredfold, thriving on stored grain, litter, and garbage. Unknown to most of us, owls keep a check on the populations of these rodents, controlling the spread of dreaded diseases like plague and rabies. It is but natural that with an increase in the numbers of their prey, the number of owlets has also increased. They can often be seen hawking moths near street lamps, catching them in mid-air with their feet and flying to a perch to finish the snack.

Spotted Owlets are not entirely nocturnal and can occasionally be seen hunting during the day. But no sooner is an owlet spotted by another bird than it is surrounded by a horde to bully and chivvy it into taking evasive action – which generally means dashing from one leafy hideout to another or diving into some natural hollow in a tree. In fact, the noise that such a collection of birds makes is a sign of the presence of either a raptor (bird of prey) or some other animal that they dislike – snakes, mongooses, or cats. If you follow the racket, you are bound to witness one of nature's most splendid dramas – the interaction of predator and prey and the ways in which they deal with each other.

Hyderabadis have a unique name for these birds – Chakwa-Chakwi. It is believed that pairs of owlets roam over the city during the day and as night descends they come to favourite perches where they narrate their experiences to each other! Our own gossiping not being enough, we would like to believe that owlets gossip too! A quaint idea indeed!

16

A Coppersmith in a Tree

You will hear him before you spot him. Let your ears guide
your eyes, and you will see him on the topmost branch of a tree,
outlined against the sky. But careful,
he is small enough to sometimes deceive
your roving eye. You imagine him to
be an extension of the branch he
is perched on, but a movement
makes you pause. Only then do
you notice the agitated nature of
a Coppersmith's method of calling.
With each *tuk* that is *wrenched* from
his throat, his entire body pivots from
one side to the other. Thus does he broadcast
his presence to the entire world! His
persistent *tuk-tuk-tuk-tuk-tuk-tuk* goes
on monotonously for minutes on end, and
you begin to wonder when he'll run
out of breath! The sound of his call is

responsible for the name Coppersmith, for it resembles the repeated ring of a hammer on molten metal that greets visitors to a coppersmith's smithy!

The Crimson-breasted Barbet, as he is otherwise called, is the most widespread member, in India, of a large family of fruit-eating, arboreal birds found throughout the tropics. Predominantly green in colour, barbets have thick, strong beaks, and short rounded wings. The Coppersmith has splashes of bright red and yellow around its face that makes it both attractive and comical at the same time. They are also partial towards figs, and one may see a large number gorging themselves on copiously fruiting large banyan trees in the city. Coppersmiths also chase insects if they are easily available, though it's a clumsy pursuit. They are fond of sunning themselves on favourite bare branches in the early mornings. They are strong and direct fliers, occasionally dipping in mid-flight like woodpeckers.

Like most barbets, the Coppersmith will dig out its nest chamber only in *dead* wood. It selects either a dead branch on a live tree or one on a dead tree and starts to dig a perfect, round hole with its beak. The chosen branch generally slopes upwards, and the bird excavates its hole on the lower side of the branch, perhaps to protect its nest from rain. A 25–80-centimetre tunnel is gouged out, at the end of which, the eggs are laid in a small chamber. I have seen a bird carry beakfuls of wood chips out of the nest, the detritus of its toil, and drop them at a distance so that traces of its work do not draw predatory marauders to its home!

If you've kept constant vigil at a nest, you'll be rewarded with the sight of the 'proud' parents bringing beaks crammed with pulpy fruit to their hungry young. They land on the rim of the entrance and stuff the nourishment down gaping beaks. When the chicks fledge, they do not have the red patches that their parents sport – these are acquired after some time. If undisturbed, Coppersmiths reuse their nest chambers for subsequent nests as well. Often, they will roost inside them at night too. You will see them early in the morning, surveying the newly lit world from their own window, before flying out to begin another day.

17

'Did You Do It?'

For some birds, the Red-wattled Lapwing is a spoilsport. For other meeker birds, it is also an avian fire alarm, a spy satellite, spotting danger from afar – a saviour. No sooner does its landscape change from the familiar to the unfamiliar, with the entry of dog, man, tiger or anything that arouses suspicion, than its strident alarm call rings out – *did-you-do-it? did-you-do-it?* Dog, man, hawk, cattle, snake, mongoose, tiger, etc., are harried and hounded by the persistent call of the bird and, harassed endlessly, they move away so that there is peace! In the time that the source of danger realizes it is discovered, other birds make their getaway. If the trespasser does not

heed the warning calls of the lapwing, the bird will dive at him repeatedly, rising sharply at the last moment, only to turn around and swoop again. The calls and the dives are an unnerving combination by any standard of aggression and ultimately have the desired effect!

Red-wattled Lapwings, so named because of the flap of red skin (wattle) in front of their eyes, have long, pale yellow legs and a quick mincing gait. Besides the wattle, they also have red around their eyes and on the beak. Their head, nape, throat, and chest are black. White washes the sides of their necks, continuing to the belly. The wings, when closed, are a dull bronze-brown. In flight, however, they have a striking pattern of black, white, and brown.

These birds are found in open grassy areas, generally close to water. They do not flock in large numbers, but three to four may be seen together. If the area is big enough, one may see many birds separated by invisible lines, each within its own territory that they patrol with great alertness. They feed on small insects, tipping forward their bodies on stiff unbent legs to pick up morsels from the ground.

The nest of the Red-wattled Lapwing is a marvel of camouflage. The female lays four eggs in the heat of summer in a scrape or natural depression on dry, stony ground. Sometimes, the nests are bordered with small pebbles or goat droppings. The eggs match their surroundings to perfection. There have been instances when the agitated behaviour of these birds has indicated a nest, but the inquisitive birdwatcher has failed to find it! In fact, their camouflage is so perfect that people have inadvertently stepped on the eggs because they were invisible!

If the day becomes very hot, a parent will fly to water where it will soak its chest and belly and return to incubate with moist feathers to cool the eggs. In recent years, there have been observations of nests on the flat rooftops of buildings in urban areas. Is this astonishing adaptation the result of reduced open spaces in cities, or perhaps a safeguard from increasing predation by dogs and cats? The precocious chicks of these lapwings are also well camouflaged. But if the parents feel that they are in danger from some predator, they have a neat trick up their wing! One of them will act as though it has injured its wing and cannot fly. It will cry out loudly to attract attention to itself, fluttering about with a half-open wing, moving away from where its young are huddled. The broken-wing ruse always works! The predator follows this larger mouthful that looks like easy prey. When they have left the young at a safe distance, the pantomiming adult suddenly becomes healthy and flies away, leaving behind a totally confused predator!

18

A Plague of Pigeons

It is difficult to imagine how the iconic symbol of peace, the dove, metamorphoses into an emissary of death. But that's just what's happening in our burgeoning, congested urban areas. This is not some Hitchcockian vendetta; indeed, the birds are innocent and ignorant, but they have help – from us.

I am talking of the apparently innocent, slyly endearing feral Rock Pigeon, aka kabootar, or pavuram (in Telugu). Before elaborating upon my accusatory beginning, let me fleetingly touch upon the antecedents of pigeon–human interactions. 'These birds originally inhabited rocks and cliffs, where they nested, and eked out their lives and their association with mankind has evolved over aeons. They've been a part of the weave of worldwide civilizations and cultures.'[17]

With the formation of towns and cities, they became our commensals, adapting to the facilities that our clustered environments provided them. It was a more involved relationship in the past when we used them for sport and pastime, for service (pigeon post), and as food. With the advance of our civilization and the diminution of leisure, kabootarbazi (the sport of pigeon-racing) has all but disappeared, technological advance in communications has retrenched the messenger pigeons, and the poultry industry has removed the roast squab from our menus.

It was our spiritual inclinations that fanned expansive pigeon colonization. The Prophet Muhammad (570–632 CE) is thought to have received divine messages from a dove sitting on his shoulder; the pigeon/dove is mentioned in the Rig Veda (1500–2000 BC). For the Sikhs, it is a bird of peace, always depicted with Guru Gobind Singh, symbolizing amity and peace in fractious times. And had the dove not returned at eventide with a fresh sprig in its beak, poor Noah would have had a hard time.[18] But rationality has seldom been religion's strength. The nuance and symbolism of religion and history have become gospel, and the human acts related to them, even if irrational, gain the sanction of communities at large and of the faithful in particular. In a world torn apart by sectarian strife, we 'worship' this symbol of peace, pandering to a mere icon and ignoring the message in real-life situations. Such misguided notions are entrenched in communities, and they congeal into deceptive dogma – blind faith can be a Trojan horse. The explosion of feral rock pigeon populations in our cities is a direct result of this conundrum.

The compassionate act of feeding birds has become an overzealous absurdity that supposedly imparts spiritually elevated feelings of benevolence. In our religious fervour, we feed feral pigeons vast quantities of grain. We strew it charitably in public places and enjoy a sense of well-being from this act of 'piety', but spilt grain brings neither peace nor divine benevolence – only huge quantities of poop from booming pigeon populations! No scripture desires the faithful to feed pigeons en masse, though they warn of the folly of excess, which we conveniently forget.

Meanwhile, the birds have become emboldened beyond redemption. Their Pavlovian instincts are easy to train. With an unending supply of grain and water, they need only breed and perpetuate their kind. Our beehive-like urban agglomerations provide a surfeit of nesting places where pigeons breed unhindered. Their traditional predators, the Peregrine Falcons, are migrants to India, preferring to patrol coastal areas, and few come inland. Their resident cousin, the Shaheen, is a forest dweller. Feral cats are simply too few and have easier food available to them from our garbage to bother expending energy to hunt pigeons. Thus, feral pigeons prosper in this heavenly urbania.

The Problem

The problem with feral pigeons is that they produce major antigens from their droppings, feathers, and blood that cause diseases in humans. This is a known historical fact amongst pigeon fanciers, those who indulged in kabootarbazi. People who have dovecotes, who race pigeons, or once used them to carry messages, are prone to the malady called pigeon fancier's lung, a pulmonary condition caused by airborne exposure to avian antigens. Pigeons, along with other birds like parakeets and ducks, also cause a number of other zoonotic diseases: histoplasmosis (caused by the airborne fungus *Histoplasma capsulatum*, which grows in the bird's droppings), cryptococcosis (another fungus, *Cryptococcus neoformans*), psittacosis, toxoplasmosis, etc. When pigeons are concentrated in large numbers and live in close proximity to us, the chance of infection increases alarmingly.

The main concern about the relationship between feral pigeons and human health is the bird's unrestricted population explosion in congested urban areas. We pamper them with food and water, we accept them as neighbours, tolerate their droppings, their feathers, the mess of their nests, and their vocalizations. We accept all this because living in unsanitary surroundings does not seem to bother us – we are a tolerant lot. As long as the garbage is outside our homes and compounds, it's not our problem. We don't also complain to civic authorities, for the result of that is, invariably, a series of unwanted headaches, and not the desired remedy. We also tend to ignore the fact that epidemics explode in congested areas in which the causative antigen-producing agent is also included.

The Disease

In recent years it has been established that pigeons are one of the causes of a somewhat mysterious lung disease, hypersensitive pneumonitis (HP), which is one of the several forms of pulmonary fibrosis. HP suggests itself on a chest X-ray and produces a distinctive pattern on a CT scan. The threat is so serious that pulmonologists frequently advise their patients to rid their environs of pigeons or even change neighbourhoods!

To be fair to the bird, there are nearly 150 other identified causes for pulmonary fibrosis related, in one way or another, largely to mouldy substances or the presence of dust in various forms of work. Contaminated water, extremely humid work conditions, and various chemicals are also suspects. Given this variety of causes, implicating the pigeon may prove difficult and indeed, it may seem a tad extreme to some people. Yet, it cannot be denied that the pigeon is one of the commonest offenders in the propagation of HP, and doctors find it difficult to ignore such a patently obvious aetiological agent, especially one that can be 'eliminated' easily.

We don't yet know why pulmonary fibrosis strikes, and we don't know yet how to cure it. Until a cure is found, we need to tackle what we know – whether this involves giving up some habits, curbing irrational pseudo-religious activities, or reducing unnaturally burgeoning wild animal / bird populations. Our addictions, even to vague religious sentiments, make us overwhelmingly selfish. We crave satiety at any cost until the addiction catches up with the addict – or a loved one. But the feral pigeon thrives on our gullibility.

Meanwhile, pulmonary fibrosis is making rapid and catastrophic progress in urban areas. People working in certain professions, or atmospheric conditions, are historically more prone to it. But the urban pigeon problem has now created a 'new' source of the disease, which strikes unchallenged across the spectrum of our social orders. The majority of its tragic victims remain unknown, but there are some it has felled that we have held in high esteem: Marlon Brando, Nawab Mansur Ali Khan Pataudi, Laurance Rockefeller, James Doohan, Peter Benchley, and many others.[19] Once afflicted, a patient is prescribed cortico-steroidal drugs, which have the potential to suppress the natural immune defences of the body. The soft tissue of the lungs gradually stiffens like cardboard, effectively shrinking the tissue that helps absorb life-giving oxygen into the blood.[20] Breathing becomes an ordeal for the patient. Simple actions like talking, sitting down, standing up, sleeping, turning in bed, eating, or drinking result in erratic levels of breathlessness, becoming moments of discomfort and adversity, and indeed, even fear and apprehension. Gradually, even the personal dignity of privacy in a toilet is compromised. No one should suffer this if we can help it or even mitigate it.

A Solution

All is not lost – there is at least a partial solution to the problem. We have encouraged and allowed the feral pigeon to expand its populations without bounds. It is in our hands to reduce its numbers. The remedy is simple; it involves no action and saves money all around, a win-win situation really. *Let us resolve to stop feeding pigeons*. Stop scattering grain in public places, in open spaces, on rooftops, on pavements, in compounds, and around places of worship (disease knows no religion). Let us make the feeding of pigeons in public places a punishable offence (by imposing fines) as has been done with resounding success around Trafalgar Square in London and in several other cities, including San Francisco, Venice, Albuquerque, New Jersey, Ontario, etc. Municipal commissioners of Indian cities must take action under the provisions of the law that castigates public nuisance. Maharashtra has some sensible laws that penalize the public feeding of feral pigeons.[21] Even the Supreme Court of India has censured feeding birds in public.[22]

Without food, pigeon populations will crash. The birds will not breed. They will gradually thin out and disperse into an increasingly wider geographical area. The processes of nature will take over, forcing them to forage for themselves – as it should be.

Human suffering will reduce, and lives and money will be saved – there will be no more expense on grain or sanitation, and we'll have cleaner buildings. Those with charitable inclinations might even reorient their largesse towards destitute and poor people. Let us restore this fallen dove of death to its rightful place as an icon of amity and goodwill.

19

The Birds of COVID-19

During the scourge of COVID-19, stringent lockdowns on human movement were imposed around the world to prevent the spread of the disease. The result of this unprecedented restriction was miraculous. In that quiescence, the planet revealed a side of herself that we had become blind to, its ability to heal quickly. A few days were all that were required for the air to clear up, for long-forgotten vistas to emerge, for neighbourhoods to resound with dawn choruses, and for atmospheric pollution to plummet. Stalled in their endless rat race, people rediscovered the reach of their senses.

The lockdowns imposed during the first wave of the COVID-19 pandemic shrank my neighbourhood's 3,000-acre birding patch (a national park that's just a kilometre from my home) to the concrete precincts of my apartment. Daily

rooftop perambulations, either at dawn or dusk, opened my ears and eyes to the diversity of life thriving around my aural and visual range, and the absence of traffic noise allowed me to see with my ears. One morning, while journaling to the meditative strains of raga Bhairavi, I heard the flute-like whistles of an Indian Golden Oriole, followed by its characteristic syrinx-clearing gargle. I rushed to fling open the windows, outside which a millingtonia bloomed. Within its redolent canopy of feathery leaves and clusters of trumpet-shaped white flowers I spied a male oriole – golden yellow, heavily kohled, pink-billed. *Oriolus kundoo* had arrived with his flute. I would like to believe that the raga had drawn him with its vibrations. Birdsong and Bhairavi stirred a spontaneous aural cocktail that resonated in my consciousness, and I felt myself letting go of, infinitesimally, a year full of despondency.

On a sultry evening, as dusk doused a flaming sunset, a kettle of about fifty Black Kites churned 100 feet up in the eastern sky, clumsily chasing glass-winged dragonflies, catching them in their talons. With a few shallow flaps for lift, and an awkward vertical stalling on their tails, they snatched at the stiff chitin with their feet and ate it in flight. They milled in a slow-motion gyre, held back by the unhurried floating of their prey, slowly drifting westwards, at a dipteran pace, till out of sight.

A female Shikra, bulging with muscular intent, clipped the housetops another evening, focused upon gore; half a dozen House Swifts tailed her, darting in her wake with bravado and brio, chittering their alarmed outrage at the hawk's audacious appearance during their watch. Not at all bothered by their admonitions, she folded imperceptibly into some foliage to impale a heartbeat to a branch.

One morning, I rose to find that the millingtonia had dropped all its flowers; a carpet of wilted blossoms spread below it. The once-green canopy began to sprout a delicate filigree of yellowing leaves. In a month, it was a depleted trellis. Yet, the oriole came on its neighbourhood commute, letting rip a queer screech of consternation at the sight of the leafy purdah having been drawn from its scented, caterpillar-rich arbour. I rejoiced in its glorious presence, a reassuring sign of resilient continuity from the spinning, tilted orb we all inhabit.

Another evening, I heard a faint sound which I could not pinpoint till I got up and followed my ear to the veranda, where a cricket stridulated, palpably, my heartstrings, and I revelled at this positive reaffirmation of life in the times of COVID, hoping the cocky Tailor Bird would miss it on its morning sweep of the balcony.

It has been over a year now since those calamitous days mantled the world, and though we still live in their shadow, the old patch has reopened, and I walk well-trodden trails once more with rejuvenated senses.

20

A Deeper Quest

The glorious sunshine poured across the water, illuminating the birds in stark relief and texture. We were late in getting to our birding spot, and the master of light had risen enough to warm the land and create a haze over it, which photographers curse.

We crowded a small parking lot atop the bend, fitting in four cars, and stood facing the water. All we seemed to be concerned about were the names and identities of the various species. Lifers are a joy for some. Some count numbers. But we do not care to sit and study behaviour, or even observe it and scratch rudimentary notes, like the fact that the Glossy Ibis population was being augmented by periodic arrivals of flocks of fifty-plus birds, at least three or four times, while we stood on the bund. They flew in scraggly dark lines, coming in from the northeast, silently strung in a gently undulating row of black wingbeats, descending without ado to their knee-deep penitent brethren.

With the water being so shallow, why were there Red-crested Pochards there at all? I did not see any upend or even dive and emerge. So why did they 'waste' their time here? Was there a deep patch of water that suited their biology that we had missed? Perhaps a day of watching would have answered that question.

I did not even look around me, from the bund, to record the landscape I was in – an unforgivable lapse. What lay on the other side of the bund? Fields, no doubt. But of what and in what state? Standing crops or harvested ones? If the level of the water in Anna Sagar was any guide, then they'd have been well irrigated, either ready for harvesting or recently cut. The water seemed to have been spread over a fourth of the full-tank level. But this was remembered in hindsight. A conscious on-site assessment, however anecdotal, was missing.

Because of some field experience, and a sketchy knowledge gleaned from reading, one surmised a patchy biology of the birds comprising the landscape. But isn't a deeper need to understand, or at least observe, one's surroundings owed to the birds, if one wished a closer kinship?

Identifying creatures and naming them brought me no closer to them than identifying and naming sports stars. At least I knew a part of the latter's ecology, but I cannot fool myself about my ignorance of the former.

What sustained the two species of shrikes that hunted amidst the thorny screen of acacias? Why did a Brainfever Bird falcon in and drop upon a low branch of a babool out of nowhere? Were the magnificent Short-toed Eagles courting airily overhead, softly whistling their amour? And those fluttering, scampering warblers? What were they chasing down amidst the cracked bark and needle-sharp thorns? So engrossed, so busy, but were they satiated? And what did they prey upon? Go deeper in your queries next time, birder.

Birding

21

A Bouquet of Benishaan

Benishaan mangoes will always have a flavour in my life that is steeped in the dry heat of the forests of Warangal (Telangana). When I sit before a bowlfull of this rich, aromatic fruit, its opulent bouquet takes me back to that summer in 1982 when eight of us had gone to the Eturnagaram Reserve Forest. It was hot – very hot. On our forays into the forest, we would pile into a small Willis four-wheel drive chauffeured by a short, rustic man full of wry wit, but with enough patience and bonhomie to survive the ordeal of driving our zany bunch around in the tinder-dry jungles.

It was beedi-leaf season, and most of the forest's creatures were disturbed by the inflow of hundreds of humans, brought in on daily wages, to collect the leaf smoked as beedis all over India. In search of wilderness, we had come to this forest, like long-weary desert travellers in quest of an oasis. The romance of being 'on safari' gave us unbelievable stamina while tramping through ankle-deep leaf litter, dry enough to make our steps sound like a chugging train; through mazes of thorny bushes whose sharpened defences seemed to have atrophied in the heat and parchedness of this land into small, bitter curves of resistance; through the middle of tiny, damp-floored creeks whose soddenness could not evaporate through the almost grotesquely green branches overhead. These were so thick and intermeshed that only chinks of sunlight fought their way to reach the numberless shadows below, making it a permanently dim dawn underneath them.

We walked through odoriferous vegetation, brushing aside generations of webs woven to trap the gossamer bodies of arthropods, wary of branches that might whip an eye. When bent double in places, to wend through the leafy tunnels created by mammals, the impenetrable green mansion above us seemed to smother us under its airy, humid weight, and we longed for a clearing and open sky! But every breath, thankfully, was laced with a zest for wildness. The creatures we encountered, whether familiar to us, or not, were deified and burnt into our minds.

One day, we walked, waded, paddled, and dawdled through the cool waters of the Dayyam Vagu, one of the shallow, seasonal streams of Eturnagaram. Hopping gingerly from sandbank to sandbank, like city dwellers jumping over road puddles, we soon realized the futility of it all and simply took off our footwear. Pleasure is walking through calf-deep water, in 40 degrees Celsius weather, stopping now and then to immerse cupped hands into the liquid and splash it over one's head, which, in that weather, was hot enough to cook the brain within! At times, the walk became palpably purposeless. We kept at it unconsciously, subconsciously aware of a hidden pursuit. Then, someone saw the magical glide of the Indian Giant Squirrels. Stunned momentarily, we stood rooted, while the water flowed around our ankles, gently reminding us of the natural history we were seeking. No swinger on ropes within a big top can stir joy so spontaneously in my heart as did those squirrels that afternoon. The mundaneness of daily life disintegrates into motes of nothingness in front of a facile, gliding squirrel! How unaware, how unconscious of our awe, how miraculously distant from our petty troubles! No poet can describe its simplicity, its perfect, arcing, well-oiled falls from tree to tree!

These forays into the 'unknown' would last all day. We would cover 20 to 25 kilometres at times! Humidity, like invisible rain, made our lives miserable, and the nights were one big sweltering discomfort. Mosquitoes necessitated the use of repellent, which made the skin clammy. And then there was the total dearth of breeze, as though we were stuck in the doldrums. Discomfort like this often drove me out of bed. But I could never make myself go beyond the wash of yellow light into the darkness – not after the sudden chill I had felt when one night, walking down the path from 'our' hut to the one in which B and M were staying, my feet in flip-flops, when I stopped mid-stride and switched on the torch. There, between my feet, as though caught in the act of some skulduggery, was a 3-inch-long black scorpion, waving its tail overhead like a flag. All pretensions of field natural history vanished and within the beat of that second,

I was desperate to be an ordinary armchair naturalist! The torch thenceforth was kept switched on, moon or no moon. I never mustered enough courage to venture into the realm of darkness, so sharply defined by the harshness of the bare electric bulbs. Prominent among my inhibitions must certainly have been some incomprehensible fear. My heart would flutter when I scampered to the dining room to wake up one of the boys for bed tea. But why did I get up in those infant hours? Was it because I was an early morning baby and the newness of the world which surrounded me, so drastically different from the soupy comfort of my mother's womb, still knocks within my breast a beat that hearkens me to rise early in strange places?

There were times during the day when everything became dead quiet, but not at dawn. A dawn chorus lifted the curtain of darkness over the forest with its enthusiasm. Bubbles of sound burst forth through the shredding blackness like an aural fanfare heralding dawn. How many people who throng to music concerts have risen at the break of day to listen to the greatest orchestra on Earth? The spontaneity of this earth poetry needs neither audience nor applause. An end in itself, it is audible to those who comprehend its perfect symphony. Have you ever heard the Koel utter a false note? Have you ever felt the flow of a Magpie Robin's song falter? Does not the stentorian Sarus sound like the clash of cymbals of the very Earth? Does not the screech of a parakeet embed itself in your heart? I used to wake up into this celebration and listen enraptured as wisps of darkness melted away all around me.

And believe me, it is no different in the city. Wake up early one morning, before the rays of a rising sun have brought the primal blush to the sky, and listen to the birdsong. There will be fewer musicians, no doubt, but the notes will be neither less true, nor less heart stopping. The Koel will hold forth in summer and often the dumpy, staidly dressed 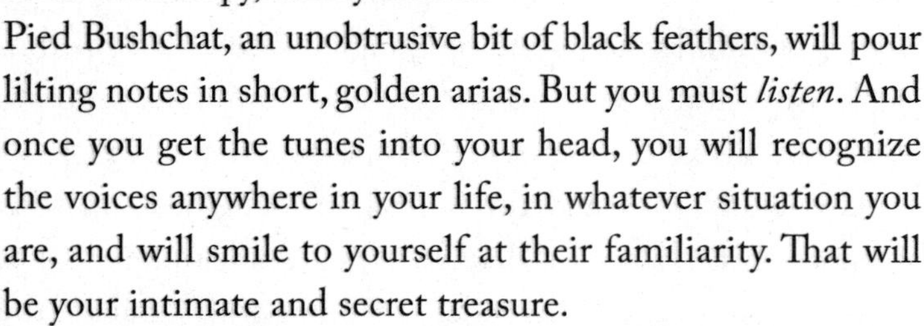 Pied Bushchat, an unobtrusive bit of black feathers, will pour lilting notes in short, golden arias. But you must *listen*. And once you get the tunes into your head, you will recognize the voices anywhere in your life, in whatever situation you are, and will smile to yourself at their familiarity. That will be your intimate and secret treasure.

At Eturnagaram, the call of one family of birds always struck me. The cuckoos. Whether it was the bisyllabic call of the familiar Koel, that swank, urbanized, prankster, or the ventriloquial utterances of the others: the Indian Cuckoo, Common Hawk Cuckoo, Plaintive Cuckoo, and Pied-crested Cuckoo. Their penetrating, persistent calls, like the seconds hand of Earth's clock, would echo, reverberate, rebound and filter through the woods all day like long-distance calls, perfected over the ages. Wherever we tramped, these birds seemed to follow with their song lines, remaining elusive themselves. Hard-headed persistence in following one call led us to a pathetically drab, indistinct splotch of grey, white and black spied inside a canopy; the very air seemed to thrum around it, with the silence of the woods magnificently amplifying its trisyllabic call. An Indian Cuckoo!

The character of a place, its ambience, takes on the sheen of the temporal moods and perceptions of the observer. In Eturnagaram, the bouquet of benishaan, rising from a bowl in front of me, overwhelmed my senses at mealtimes and mingled magically with the deceptive quiescence of life deep inside the forest. The heat of the day was an invisible, all-pervading presence that continued into the still night, when the temperature dropped marginally and grudged us an uneasy somnolence, despite the jungle reverberating within its leafy confines with repetitive, night-amplified sounds. There was also a distinct sense of sukoon (tranquility).

Come any May and the inevitable mango season with its sun-warmed aromas, that Eturnagaram summer flashes in my mind's eye, and I wonder – can one truly evaluate the usefulness of a fruit?

22

One, Two, Three … Ten, Twenty, Thirty!

The Asian Waterbird Census was initiated on the Indian subcontinent in 1987 and has since grown rapidly to cover major regions of Asia, from Afghanistan eastwards to Japan, Southeast Asia, and Australasia. The census thus covers the entire East Asian–Australasian Flyway and a large part of the Central Asian Flyway … Till date, more than 6,100 sites across 27 countries have been covered with active participation of thousands of volunteers.
– Wetlands International[23]

They said I would be participating in a scientific experiment being carried out throughout the Indian subcontinent. They said it would be meaningful, purposeful birding, geared towards a finish line they called 'scientific ornithology'. They said that in all of Andhra Pradesh there were only a few enlightened folks who were engaged in this essential, this phenomenal task. Imagine your good luck, they said, on being one of them! 'You will be taking part in a nation-wide census of wetland avian populations. Over the years, your contribution will be a part of the history of cooperative ornithology in Asia!' they said.

I fluffed up my breast feathers and rocked forward and backward with the tightness of an unknown pride inside my body, perched on the ordinary, everyday branch of 'just birdwatching'. And crooning to myself a subsong, I readied my untried wings for the first flight into the azure realms of scientific ornithology.

The road to Palmakole Tank is a good one for it is 30 kilometres out on NH 7 (now NH 44) and I got there by 7.45 a.m. They had mentioned that waterfowl would be best spotted after 8 a.m. To reach the tank, one must take a kuchcha track cutting away to the left just as you enter the town. It takes you through a lot of ups and downs, minor ones in comparison to those I faced later, for a couple of kilometres. Suddenly, you see the water through the trees, out-of-focus pinpoints of sunlight reflected on the wavelets. As I approached, I realized that the water spread was not at its optimum; the past three years had seen a drought, and the water had dwindled. They had insisted that I would see more birds for this very reason. 'The scarcity of waterbodies would concentrate bird populations', were the very words they had uttered, and I had swelled with a constricting smugness of one who is given a crucial task.

And here before me was the water, spread over nearly

fifteen acres, held there by a man-made bund on the western side and by a general, natural gentle rising of the land all around. Around the water was semi-hard soil of suspicious elasticity, more cracked and folded than the skin of an aged bull elephant but of the same colour. The evaporating water had done the trick. This strip of land met fallow fields with the dusty brown soil that had not released much of its petrichor into falling rain for the past three years, and beyond this fallowness stood some beautiful old tamarind, neem, and babool trees. Further on rose the rounded, smoothened, quintessential boulders of the Hyderabadi countryside. Above the 'silences' that pervaded this scene came the drone of a stone crusher.

But what of the birds on the water? There were birds and there was a shikari. Three shots put the waterfowl up twice, and twice they circled the water and returned to settle. But the third shot got a flock of ducks up, and they did not return. I did not see who had fired, but then such a one is good at subterfuge and camouflage! When I finally settled down, having set up the scope and strung the binoculars from my nape, the waterfowl settled too after having stealthily put more water between themselves and me with an innocence

only they can exude. I then began the great scientific work I had come for.

I got the numbers of the larger birds right the first time round. And these were put down as so many large or smaller or little egrets, and so many Grey Herons, and so many Brahminy Duck, and so many Shovelers. This was good fun and easy arithmetic to boot. It was also an excellent job in science. Then, I looked through the binoculars at the smaller waterfowl and saw many on the water. They were of colours I could not perceive except for the hazy splashes of brightness and dullness that colours become at distances. So, I peered with a single eye through the scope, having shut out the world from the other with an open palm so I could keep that eye open but unseeing. I saw the full colours, and I was thrilled and uplifted by my ability to identify birds of different feathers. It was simple. Yes, and again I got them right the first time round. Or had I? Surely something was wrong. I was missing something which could be vitally important. And then it struck me. Great Auk! Nobody had warned me that amongst all that colourful pomposity, amongst all those identification flags that floated across the water in front of me, would also be their spouses! Dull, drab reflections of each other which made identification at that distance mind-boggling. I strained and I stared, and very soon the coloured birds began to change and go up in bright spots of orange and so many thingamajigs arose from the water that I took away my eye from the scope. My head reeled and many a fowl image danced on the water before me.

Time is the best healer they say and so, I waited. The dancing stopped, and the world spun into vision once more. Females should be censured from such censuses I thought. *I* would tell them! Only *I* know what chagrin I had grinned and fluttered through. They had me baffled, nullified, disheartened, pinioned, hamstrung. And from *that* distance! I know birding is a hobby of great patience. But this! It was a hard journey full of crosswinds and troublesome waves of heated air and predatory spots before the eye and unknown, awesome mysteries.

When I came to the waders, having dabbled in troubled waters, the going, ironically, was easier. Sexes were alike and species recognizably different. The number of *Himantopus* was not stilted and god's wit was visible on blackened tails. A large flock of *Calidris minuta* were busily doing a tap dance stint and a few of their cousins were at Temmincks'. The red and green shanks were also poking in the squelch, now and then lifting into flight, whistling hauntingly across the waters. Some sandpipers were of the common variety, some were green and some just out of the woods. A dunlin tried to maintain dignity among so many stintsmen, looking down at them over a slightly decurved bill and a statelier feeding gait.

When it was over and I returned to the fold, I asked them, 'Have you counted female ducks, piled upon a narrow strip of shore one on top of another? Have you tried to identify vague shapes through the rising hot air and not cried "Ugly duckling!" with disgust?' They quailed at the onslaught and denied having had the privilege of such experiences. 'Your talk is "cheep",' I said. 'We philloscopii should first learn to identify ourselves,' I said, 'nondescript sons of warblers that we are!' And I fumed and fretted to the extent my birdbrain allowed. Why did that quack suddenly escape me? Why did I turn pink in the face? Why did I get that pang of ostrichitis? Echoing on the eddies I had created came a confident, agnostic query. Have any of you tried counting Dabchicks? Have you experienced the confoundment these joyous avian porpoises can put one in?

I thank my stars that the next census is still many suns away!

23

End of the Day Birding

End of the day birding, or birding at dusk, has a quietude that is absent in the hustle and bustle of early morning birdwatching. It has a unique serendipity – activity is reduced, and things happen with the stateliness of a day's energy gradually subsiding. There is peace to be found in the quiet ditty of a male Pied Bushchat as he seats himself upon some subtly prominent place and practises his muted melody, while jerkily, almost nonchalantly, flicking his tail up and down. As it darkens, he swoops to his roost and calls it a day. The Shikra, who has invariably hunted and fed just before dusk – having espied an unwary squirrel kit or Garden Lizard from an elevated perch, manoeuvred through the trees (leaving a wake of anxious calls of *'ware-hawk* from flustered lesser passerines), pounced upon, caught, and waited for life to ebb away from the prey before flying to a favourite feeding post and satiating herself – has also found a resting place in the recesses of a leafy peepal.

A sisterhood of White-headed Babblers arrive by a devious, circuitous route, traversing up the spindly stems of a henna hedge, switching this way and that through the leaf-shadowed depths of a small mango tree. They float into their roosting flame of the forest tree on an invisible thread of contact calls, alighting one by one upon a branch like stitches upon the tree's fabric. Having arrived, they huddle up to each other, some facing this way, some the other, and settle down for the night. A Golden-backed Woodpecker sends a peal of manic laughter ricocheting through the warming, coppery, tinder dry garden. A Coucal booms its retirement for the day, and a drongo traipses behind crepuscular midges. From a third-floor apartment in the dead fishtail palm come the first chuckles and coughs of just-roused Spotted Owlets. I head home, content that *life* does not cease just because darkness blankets the landscape, but a nocturnal fauna and flora take over from the diurnal.

24

Bird on a Wire

Since the world got wired, birds have found a new perch. Watching birds on a wire is a fascinating pastime. I have stopped innumerable times on the new flyover connecting the Raj Bhavan area in Hyderabad to the Secretariat to admire the neat rows of migrant swallows strung up like musical notes across a calm blue sky. The birds are usually spaced out at wing length, preening or generally acting social. When something disturbs them or they are in play, they slip-slide away with an explosion of exuberant chittering.

One very wet and rainy day, a couple of friends and I followed the lustful, feverish yells of a Brainfever Bird in the forests of Narsapur, near Hyderabad, until we found the fellow high above the trees, lit upon a wire, singing his heart out! The wire is a perch to an untold variety of birds, and it is one of my favourite watch-posts because a bird on a wire is completely visible and much easier to identify. Not that I haven't been stumped by one and turned away, blinded by my ignorance. Why can't warblers just play hide-and-seek with leaves?

The wire is also a crow's nest for the avian sentry. It is the cool place to promenade plumes and air arias, to flirt and to fight and flee. The wire offers the ideal change of perspective for pouncing upon the scrumptious little critters scrounging within the angled folds and hinges of grass leaves. It is the launchpad for avian artists to trapeze across the azure top's arched dome, snapping up midges in mid-flight with poise and panache. The wire is a clothes line for nature's wardrobe where the fabric is a multi-hued feather.

Have you heard of the wire-perching-bird list? No? Then why don't you start one? Which birds use wires, for what, at what times, in which seasons? Get wired to wire-fancying birds. Soon enough, you'll have enough material for your story!

25

Revisiting Shamirpet Lake

Local administrative bodies must realize that strangulating waterbodies for the cause of development is doubly damaging. It not only chokes the water, that field of stars and clouds, but cheats investors of their promised scenic landscapes. Rolling back the years on a once-favoured birding haunt reveals a tale mirrored in a thousand habitats across India. The greatest danger of rapidly disappearing wilderness areas is generational amnesia.

The Winter of 1982

Shamirpet Lake, on the outskirts of Hyderabad, a small man-made lake, formed in a depression of land and bunded on one side, is a favourite birding area for the Birdwatchers' Society of Andhra Pradesh (now the Deccan Birders). Its water is used for agriculture, and boulders of every size and shape surround the waters and form islands within the lake. A few agricultural fields are on one side, amidst the rocks,

117

and near the water's edge, some villagers are making bricks from the wet mud. A lone fisherman punts his makeshift raft near the shore. It is a vista of peace, quietude, and ascending lark song. At every step, pipits, larks, wagtails, and Little Ringed Plovers dart away or rise if pressed, settling again a few yards away on the close-cropped green turf. The squeaky-clean air and the angle of light on the entire landscape enlivens it into a graphic three-dimensional kaleidoscope of such exquisite detail and vitality that one could not wish for more perfect surroundings. Small flocks of Grey-necked Buntings, perhaps at the southernmost outpost of their migrant wanderings, browse amidst the boulders strewn all over, like marbles by a giant hand. A pair of Indian Coursers saunter in the fields some distance away. On a boulder shaped like a platform in the middle of the lake is a flock of nearly eighty Small Indian Pratincoles, sunning their milky bellies. The list of birds crosses seventy species before noon. A dozen species each of waterfowl and waders (including Curlew, Black-tailed Godwit, Curlew Sandpiper, and Dunlin), at least twelve to fourteen species of raptors (Osprey, Peregrine Falcon, Red-headed Merlin!), an abundance of larks (the endemic Syke's Crested Lark is a guarantee!), pipits (at least four species!), and a variety of other perching birds are among the avian attractions of this wonderful place.

The Winter of 1996

Heavy rains have filled the lake. A large, ugly building looms on the opposite (from the Deer Park) shore, housing a speciality hospital. A tourist complex promoted by the Andhra Pradesh Tourism Department stands, almost forlorn, adjacent to the park. Omnipresent is the nerve-racking sound of breaking granite. A monstrous, unfriendly neighbourhood stone crusher hammers the dirge of progress into a numbed mind. The entire landscape is dotted with workers busy hammering chisels into boulders (even the small 2x2 inch ones are fair game!), an occasional blast of dynamite (spelling the doom of yet another rock) perhaps a shot in the arm. The dry topsoil rises in eddies at the slightest breeze, and the ground is littered with granite chips. Fishing net floats dot the lake as half a dozen fishermen are busy at various points over the water. One wonders about the plight of the poor tourist! And the birdlife? Gone are the buntings, gone the coursers and the pratincoles (their island rock sliced and broken into scarred ugliness one summer when the water had dried up). Your ears strain for lark song, from a bird that Wordsworth christened 'ethereal minstrel',[24] and Shelley heard as 'unpremeditated art'.[25] The flotilla of the waterfowl is thin, and shorebirds slip away in hurried flight. Nature clings on to a carpet that is disappearing rapidly from under its feet. A pair of Ashy-crowned Finch Larks

are busy raising a brood of three chicks in the shelter of a small boulder, in a perfectly camouflaged nest, a few yards away from a man busy breaking stone.

The Future

The very existence of this once-beautiful habitat is at stake. The threat is from blind and ignorant development, and if the surroundings are cleared of all boulders, can the flora survive for long? This deterioration is so widespread that it almost certainly encircles the entire catchment area of the lake within its tentacles. The development of the area is smothering the lake; as the topsoil erodes, agriculture suffers; as the water table dries up, triggering a local water crisis, the fishermen suffer. Meanwhile, cattle and goats suffer for lack of browse, adversely affecting those depending on them for their livelihoods. Would a tourist like to visit such a place?

Only if all these activities are stopped within the catchment area of the lake, right up to its shores, will Shamirpet survive in its old glory. It is imperative that developers take cognizance of the role of natural landscape in an area demarcated for development. Birds (through a record of their presence and absence over time) are indicators of the quality of the habitat and an alarm system that alerts us. Lark song is one of the cornerstones of sustainable development, and it is the presence of the latter that ensures the satisfaction of tourists, as well as sufficient supplies of water for agriculture and drinking.

26

Remembrances of Birdsongs Past

Having exhausted all excuses to postpone escorting my wife and kids to the handicrafts fair at Shilparamam in Hi-Tech City in Hyderabad, the day finally arrived when I was cornered into being their chauffeur and chaperone to this unpalatable melange of noisy crowds, dusty paths, blinding lights and blaring sound. Though I could have refused and risked a week-long cold war, the pools of endearing expectation in my children's eyes made me acquiesce. I agreed to go along on the condition that they leave me to myself once we reached the place. We arrived at Shilparamam about half an hour before sunset, an hour when it was bright enough for them to verify and match the colour of garments under natural light. We agreed to rendezvous back at the gate around 7 p.m. Shopping instincts aflame, they stumbled out of the car and raced towards the row of stalls.

The fair itself had everything I loathed except that it was picturesquely surrounded by boulders of innumerable sizes and shapes, piled and scattered all over the place, breaking up the landscape with the immensely pleasing contours that rock formations have. *If nothing else*, I thought, as I wandered listlessly through the pressing crowds, *I could climb on to a sun-warmed rock and watch the play of firelight from a setting sun change the texture of the land.* I made my way to a boulder on the western edge of the grounds and sat comfortably on it, revelling in its warmth. I let my eyes wander along the land as it fell away into a shallow valley with remnants of paddy cultivation on its floor. There had been enough water here long ago, to support a paddy crop!

Suddenly, above the clamour of the fair, I heard the muted ditty of a Pied Bushchat. He was perched on a small sitaphal tree, nervously twitching his tail at this intrusion of privacy. I was still and silent for a while and he too let the serenity of the hour prevail. Gradually, the tumult of civilization receded from my consciousness as I watched the shadows lengthen away from the rocks and the sharp edges of grass blades softened into a fringe of fine, coppery fuzz. Other birds began calling and singing. Red-vented Bulbuls fussed for perches in lantana bushes, Indian Robins stood each other off at territorial boundaries with cocked tails and flashing white epaulettes, Small Green Bee-eaters perched on rocks twittering to each other, and from afar came the tobacco-roughened rasp of a Painted Partridge – *subhan-teri-kudrat*. Not to be outdone, an Iora whistled his lilting onomatopoeic *shaubheegi* from within a neem tree in the darkening valley below.

A bird's song can fill space with such simple music that one is compelled to cease all activity and simply stand and listen. Even as I sat there, meditatively absorbing this palpable natural beauty, my mind sauntered away into the recesses of my memory. Once again, I heard the dawn chorus of the birds in the Borivali National Park (now Sanjay Gandhi National Park) emanating from the depths of the valley, like the notes of an organ filling up a cathedral. Once again, I heard the fluty serenade of a Himalayan Whistling Thrush's exaltation in the hushed darkness below a balcony in Mussoorie as he searched for crumbs from the previous day. I heard the song of the White-throated Ground Thrush as he sang lustily from a sun-dappled gully in Wellington as my wife and I sat, missing out on lunch, mesmerized by that pure, joyous cascade of avian music. I heard the ventriloquial *one-more-bottle* of an Indian Cuckoo on a sweltering afternoon in the Eturnagaram forests of Warangal as my friends and I tried to find the source of the song! I heard the ludicrous jugalbandi of a pair of Peninsular Scimitar Babblers, as they performed at eye level at a coffee estate in Coorg, throwing their heads about in the fervour of their performance. I heard that joy of urban gardens, the Magpie Robin, the worthy competitor of the legendary Nightingale, and the hesitant, husky notes of the beautiful, dumpy little Tickell's Blue Flycatcher, singing in my own patch of green at home. All these songs rushed at me as I sat listening to my memory speak, watching the sun sink below the boulders on the opposite ridge.

The dust had not yet settled when the first nightjar began to call, hesitantly at first and then gaining confidence with each succeeding volley. Within no time, many more had joined into a chorus which pulsed and throbbed in the darkening air with potent meaning. As I got up to leave, the nightjars were still calling. I knew they would continue to almost till dawn when they would finally fall silent. It has been thus down the ages. Now, they are being pushed farther and farther away by our expanding urban jungles where there are no broken grounds for them to hide. Instead, erstwhile habitats have been overrun with concrete pavements and roads which do not allow rain to percolate to the water table. Such water now stagnates in puddles, forming cesspools and breeding grounds for the nightjar's favourite food, those vectors of disease increasingly resistant to every chemical developed for their annihilation, the mosquitoes. We cannot control them, and we will not allow those who can to either.

There is something so remarkably elemental about the call of the nightjar that it seems as though the very Earth is speaking. I realized then that it was indeed the Earth speaking, reminding me that the natural heritage which lay around me was the inspiration for all that was going on over the hill under the banner of cultural heritage. In our anxiety to applaud and deify our cultural heritage, can we afford to forsake and forget the splendour of our natural heritage?

27

Uma Maheshwaram: Sacred Indeed

It is a 150-kilometre drive to Uma Maheshwaram and a halt midway for breakfast at Amangal is inevitable in chilly November.[26] Hot idlis off a roadside pushcart, piping hot cuppas at the omnipresent Udupi hotel and freshly plucked fruit from fruit vendors' overflowing baskets! One vehicle actually stopped at a stall for sizzling hot pooris and aloo sabzi! You see, birding day delights can be gastronomic too! Fortified and warmed, we piled in for the first stop – the Dindi Project.

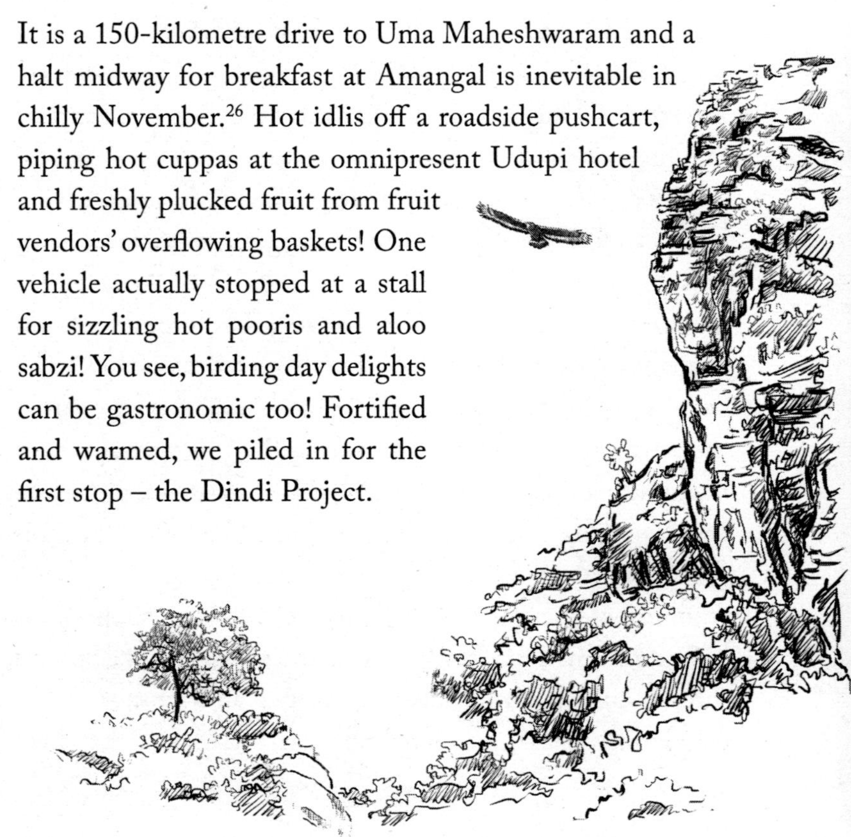

Here, a dam has been constructed across the Dindi, a tributary of the Krishna. In years past, this waterbody has hosted flocks of 2,000-plus Greater Flamingos. This year's abundant rain topped up the reservoir, spreading a glistening sheet of water far and wide. From the dam we cannot spot anything exciting. A few gulls float in the distance on its still surface. A pair of nimble River Terns twist and dive repeatedly in a shower of sparkling droplets. An adult Brahminy Kite glides desultorily overhead, its silhouette confusing some. As if to reassure them, it dips a wingtip and slips across the sky where the sunlight bounces off its clean, white breast and chestnut plumage. On the distant shore, a few trees shine with large, glistening white fruit. Egrets! Past the causeway, near the Fisheries Department buildings, we spot our guide, a forest guard. A giant peepal on the property is home to a large colony of Indian Flying Foxes, a cloaked judiciary, ruminating jurisprudence, in an upside-down world!

Uma Maheshwaram is a sacred grove around an old Mahadev temple. The shrine is constructed on a thin ledge running along the side of a hill that rears 200 metres from the surrounding plains (approximately 500 metres above mean sea level) to form the Amrabad plateau, a part of the Nagarjunasagar-Srisailam Wildlife Sanctuary. At this spot, the plateau forms a C-shaped amphitheatre that has a stunning scenic impact when viewed from the plain. These hills are formed of lateritic soils and support dry deciduous forests. Uma Maheshwaram is clothed in a unique, moist deciduous flora as perennial streams, that leach the sheer rock above the temple, feed it. During monsoon, these rills transform into a curtain of waterfalls that generate enough humidity and moisture to sustain bryophytic flora amidst their musical splashing. Moss carpets moisten unused surfaces in green velvet. Radiating sprays of ferns abound in cracks and crevices, rooted to and sustained by thumbnail crescents of soil. Various shrubs and trees cling precariously to the rock wall, spreading a web of roots on its surface, indicating soils of extremely poor nutritive value. But they are sustained by the damp that percolates and oozes from the rock. Various Ficus species let down thin aerial roots from the rock wall rising above us in their search for terra firma. Camel's Foot Climber, that giant lianous creeper, smothers tree crowns in fecund abandon. Its leaves are large enough to be used as dinner plates by local inhabitants. When it flowers from February to March, what a glorious sight it is!

Looking at the tor above the temple, VT ponders aloud, 'How long will this stony shoulder survive the ravages of natural and human-induced erosion?' Below, the land angles to the plains, thick with vegetation, in leaf-heavy canopy. A recent study of the biodiversity of sacred groves in Andhra Pradesh[27] revealed that Uma Maheshwaram 'has the richest flora in Mahbubnagar district. About 400 plants were recorded during floristic studies of which about 150 are perennial. Bryophytes like *Riceella*, *Notothallus*, Moss, *Marchantia* and Pteridophytes like *Salaginella* spp., *Pteridium* spp., *Adiantum incisum*, *Actiniopteris*, *Petris*, etc., were observed ... It is an important grove requiring better protection.'

Sadly, 11.30 a.m. is not a very good time for birding, but we have no choice. We have reached at that hour and must make the most of it! While most drive straight up to the temple, a few park at the foothill and walk up the winding 1.4-kilometre road. White-browed Bulbuls' electric-spark warbles follow our footsteps. They always try to say more than time permits! Jungle Babblers abuse and curse in guttural monosyllables from the undergrowth. A hidden woodpecker's drumming echoes in short staccato bursts, ridiculing our bird-spotting prowess. However, the distant sound of a Large Green Barbet's *kutroo–kutroo–kutroo* augurs a fruitful day ahead.

As we reach the top, having walked up with friends, a few birdwatchers are waiting near the car park, while some have gone ahead, led by the effervescent MSK. Bonnet Macaques throng the place. Many females have young clinging to their bellies. Two to three troops range along the shelf, either collectively bullying pilgrims and visitors to jettison their eatables or stealthily stalking someone carrying food, biding their time to dash in boldly and snatch the savoury in an unguarded moment. SJ falls prey to one wily individual who grabs a loaf of bread from his lap as he sits cross-legged and sketches!

As we take in the view, the ferns are a source of wonder for some, and the vista of the valley, as it stretches away into the distance, to others. SU, relatively new to birdwatching, describes what must be a White-throated Ground Thrush, above us on a rocky ledge. While we crane our necks upwards, someone spies a bird of prey, wings swept back in a shallow glide, heading over the cliff above us. 'Bonelli's Eagle,' says R. In quick succession, we spot a buzzard, but are not able to identify the species, and two Crested Hawk-Eagles. They all use the warm air rising from the valley to conserve energy. The song of Tickell's Blue Flycatcher rises from the foliage below us. Not many are familiar with it as their auditory faculties have yet to be fine-tuned to nature's sounds.

MSK's group is ahead, tilted heads peering up the rock face, and MSK's glee is contagious. 'There are YTBs here!' he exclaims. The excitement of viewing an endemic rarity, an avian star, grips everyone, and the birdwatchers are charged with expectation. The Yellow-throated Bulbul (YTB), Blue-headed Rock Thrush, and Verditer Flycatcher are seen in quick succession. Questions, explanations, discussions, and disagreements follow as field guides are opened, notepads scribbled in, descriptions spoken, and vehement denials expressed. The Grey Drongo and its smaller, smarter, White-bellied cousin, unconcerned witnesses of this gesticulating group, sally from perches after aerial prey in thrilling dives and swoops, vying with Small Green Bee-eaters for these midges.

The YTBs, the tellurian relics of prehistoric landforms, adroit survivors of a demanding landscape, are in small groups of two to three or more, generally feeding on small berries or chasing one another through the foliage of trees that cling like limpets to the vertical rock face. Inexplicably, much to my embarrassed chagrin, two plumage types occur among these birds! One, as described in the books, the other with a yellow chin and throat, a white chest and belly, yellow thighs, ventral region, and under-tail coverts. Above, it has a yellow crown and head, olive back, and olive brown wings and tail, which is tipped yellow. The yellow head and clean white belly are bewildering. SS, who has done some work on the species, informs us that EL reported similar-looking birds from the lower Palanis, a hill range in Tamil Nadu! This raises several questions that require answers from the experts. It is a thrilling experience to watch them gambol with such abandon in front of us! Their bubbling calls are like those of their commoner cousins of the plains, White-browed Bulbuls, but slightly different, these being less garbled and at a higher pitch.

We watch House Swifts balling in the sky in loudly chittering flocks and spot Dusky Crag Martins among them. SJ settles to sketch, while others engage in light banter amidst rumbling bellies! Blue Rock Pigeons clap wings as they take off from craggy perches. I collect glittering moments of natural sounds when those of civilization cease. A multitude of liquid splashing. The snapping of a bee-eater's mandibles. A Grey Junglfowl's proclamation of its territory. The lisped *tiss-yip* of the impatient, zesty Greenish Leaf Warbler as it searches along leafy stalks for titbits. The aural delights of birdwatching!

SJ has either finished his play with lines, or gastronomic requirements have taken precedence in his scheme of things. We picnic with relish and bonhomie – undisturbed, surprisingly, by the bandar-log! For the post-lunch session, our group splits into two. One prefers to birdwatch as its members walk down the hill on the rough-hewn stairway.

The other retraces the earlier route. Dividends for the latter are almost immediate. Besides the morning's goodies, we also get a clear view of a Large Green Barbet. KCR and I walk ahead, trailing some YTBs. I peer over the railing and transfix an Emerald Dove into immobility. It is perched 12 feet away on a tree! After an initial hesitation, it rockets away along the hill, showing itself to the others too! When we have all gathered at the bottom of the hill and begin comparing notes, those who descended on foot will not be cowered by the dove. They have seen a Yellow-checked Tit! Forty-six species have been sighted today. As our exhausted group piles into cars, the collective opinion is to stay overnight at Achampet in the future and birdwatch here during proper birding hours.

28

A Bird in Hand: The Patience and Persistence of a Bird Ringer

To popularize bird banding (or bird ringing) in India, the BNHS conducts camps under the expert guidance of its legendary scientist Dr Balachandran. I attended one such camp in August 1999 in Chittoor district, Andhra Pradesh. One of the reasons to band birds is to study bird migration. Migrants are trapped in large numbers and ringed with a metal band that has the address of the banding organization embossed on it. The idea is to recover the ring, either from dead birds or when birds are trapped elsewhere. Once a bird is in the hand, a scientist records data such as breeding condition, moult, age, ectoparasites, general health of the bird, analysis of blood samples, etc. The differences in plumage between the sexes and between an adult and a juvenile are also recorded. There is another motive for banding resident birds with coloured bands (without any markings). They are used in intensive studies of the ecology of a species so that individual birds of that species are recognizable in the field through optics. A scientist records all necessary data when each bird is trapped the first time, adding to his database with each subsequent entrapment.

The White-browed Bulbul's heart beat like a metronome against my fingers. I held it, as a fast bowler would grip a cricket ball for a swinger, except that my index and middle fingers were on either side of the bird's head instead of the seam of a cricket ball. Its wings were folded against my palm and its legs clasped between my third and fourth fingers. Though it was immobile in my grip, the creature struggled with verve and belligerence in its bid to fly away. I was required to hold the bird firmly enough to restrain it from escaping but with a light touch to prevent injury so that its frailty remained intact when it was released.

This required concentration and my palms had begun to sweat slightly. Its soft body feathers were coming off in small bunches, perhaps as a means of defence akin to a gecko's disposable tail trick. I must say, the little thing had spunk. All the while I held it, it screamed blue murder in its loud tripping avian tongue and never missed an opportunity to nip my hand. Such chances it got aplenty as I constantly juggled fingers to replace a freed wing or to re-grip a squirmed-out tarsus within the temporary cage of my digits. In guilty silence, I read my anthropomorphic emotions playing in its eye and beheld in its structure such perfection of line and colour as to warrant for once the justification of holding 'a bird in the hand'. Indeed, 'a mouse is miracle enough to stagger sextillions of infidels'.[28]

That *Pycnonotus luteolus* must have been momentarily flummoxed when it was inexplicably snared mid-flight from the top of one bush to the centre of another. After a brief struggle, it lay still, suspended within the folds of a fine, black nylon mist net. It was extricated without much ado, from a seemingly hopeless entrapment, by Dr Balachandran's deft fingers and immediately placed inside a small cotton bag with a drawstring.

Leaving three other 'netted' birds to his trusted trapper Sethu, Balachandran, fondly known as Bala, led us into a clearing where we sat on the ground, and he began explaining the intricacies of bird banding.

Our group of five was attending a bird banding training programme organized by the BNHS, and we could not have asked for a better guide than Bala, for he has single-handedly ringed about 50,000 birds! 'The primary concern is for the health and safety of the bird. One must be gentle while handling it, deft in taking measurements, discerning in deciphering moult and quick to complete the task and release the bird, to minimize the creature's trauma,' he instructed. *Easier said than done*, I thought, as the bulbul in my hand ferreted out a tarsus for the nth time. But as I watched him, I noticed that the Red-whiskered Bulbul he held seemed to be a part of his hand! His movements were adroit as he examined the bird like the friendly neighbourhood doctor going over a patient.

'Fix a ring as soon as you get the bird. Sometimes, they escape while you're reading the moult score. This way, at least you've banded it.' He pulled an aluminium band off the long plastic tube that held it and placed it around the left tarsus. 'Left for odd years, right for even,' he cautioned us before pressing the band closed. 'In their enthusiasm to ring the bird, people inadvertently transfer their strength to both hands and before they realize it, end up wringing the poor thing's neck!'

All of us were doing this for the first time. Sitting around Bala, we learnt about birds as we had never done in all our years of birdwatching. To actually hold feathered bipeds in my hands and measure the lengths of their wing, bill, tarsus, and tail were practical aspects of field ornithology I had always dreamed of doing. The process of measuring a tail is a most unceremonious act. One blows apart the rump feathers to uncover the preen gland and places the starting edge of the ruler against it, smoothening the tail feathers along the metric divisions to read off a number. A final indignity to the poor creature was to blow hard on its belly and quickly gauge the condition of its brood patch while the feathers remained separated under your breath! (Any aspirations of being reborn as a bird were now fast receding!)

To read moult, a wing is swung open smoothly to reveal primary and secondary flight feathers 'and the narrowest hinge in my hand puts to scorn all machinery', said Whitman famously.[29] Moult begins from the primary nearest the bird and moves outward towards the tip of the wing, referred to as an 'inside out' moult. Some birds don't follow this pattern and the moult becomes haphazard, known as 'irregular moult'. Besides measuring and reading the moult score, we have to recognize young birds by the condition of their plumage. While one of us went through these motions – juggling ruler, bird, and parts of the bird, concentrating, and calling aloud measurements – another recorded the data on specially printed forms that were analysed and stored at the BNHS.

Geology of the Region

Geologically, the entire Tirupati area in Andhra Pradesh is awe-inspiring. Its bedrock is supposed to have been set more than 2.5 billion years ago during the Archaean period, making it one of the oldest pieces of real estate in the world, part of that vastness of antiquity, Gondwanaland, that later disintegrated into various continents, this piece being the peninsular portion of India that broke away from the eastern coast of Africa and floated northeast towards the Eurasian plate. Their monumental collision created the magnificent Himalaya. This aged landmass is E.M. Forster's incomparably evocative Dravidia, whose 'high places have been land since land began … They are older than anything in the world. No water has ever covered them, and the sun who has watched them for countless aeons may still discern in their outlines forms that were his before our globe was torn from his bosom. If flesh of the sun's flesh is to be touched anywhere, it is here, among the incredible antiquity of these hills.'[30]

The multiple hills of the Eastern Ghats complex are remnants of an antediluvian range once mightier than the Sahayadris are today. Millennia of ceaseless erosion by wind and water have reduced them to an uneven row of broken vertebrae of a spine that is now old and crumbling. Indeed, it is on this land that the truly relictual species of our subcontinent precariously reside, epitomized by that frustratingly elusive noctambulating Jerdon's Courser, whose closest cousins pace the vast plains of eastern Africa. Birdwatching, and indeed, natural history itself, acquires an exciting, luminous aura when hitched on to perspectives of geology and biogeography.

In the Plains at Bhakarapet

We were encamped at Bhakarapet village approximately 37 kilometres west of Tirupati town, in the Andhra Pradesh Forest Department's sprawling and comfortable guest house complex. The mist netting had been done in the scrub forest on the road to Tirupati, within 8 kilometres of the village. Close by, some encroachment of forestland seemed to have taken place as paddy fields and orchards had sprung up. The xerophytic scrub began 3 kilometres beyond the village. On either side of the road was a thick growth of impenetrable thorny bushes criss-crossed with cattle tracks and sprinkled with small clearings.

Bala knew this area since he had already conducted two camps here earlier in the year. Though limited in its variety of vegetation (but this could be merely our ignorance!), the diversity of bird life would surprise us constantly over the next few days. We worked with twelve to fourteen nets installed in two different areas. On the third and fourth days, we shifted one group of nets to the open area close to the paddy fields and orchards. We ringed 319 individuals of 41 species during our stay. The most common birds ringed were the White-browed Bulbul (41) from the scrub and the Baya (38) from the paddy fields. We had placed a net close to an active Baya colony, and as we became adept in the art of handling birds through the various steps of banding and release, the challenge shifted to trying to identify the species inside a cotton bag before it was taken out. Our clues were the weight of the bag, size of the bird and above all, those epithets of anger, outrage or plain fear that emanated from within.

Some species were given to an effusion of the choicest avian expletives, which gave them away quite easily. White-browed Bulbuls, Jungle Babblers, Bayas, Blossom-headed Parakeets, and Common Woodshrikes belonged to this set. With those that were quieter, it was mostly a game of chance. But as each one emerged from its bag, our unabashed wonder at its hues, so brilliant at times as to make one gasp in disbelief, and at the delicate perfection of its form, always left us with a feeling of elation and enrichment. We felt privileged to have held *and* released with our own hands these shining fragments of eternity.

In the Foothills of Talakona

For a change of scene, we went to Talakona for two days. This is a block of reserve forest inside the 506 square kilometres of the Sri Venkateswara National Park. Our destination was a sacred grove around the Sri Siddheswaraswamy Temple, which lies 3 kilometres downstream of the Bugga Vagu reservoir, which was constructed on the eponymous river ('vagu' in Telugu). Pilgrims visit the Talakona waterfall, as it is known, to cleanse themselves of their sins. A fold in the escarpment of the Palkonda Range, overlooking this temple, had resulted in a narrow valley that had been gouged by the Bugga Vagu. Birdwatching was phenomenal along this stretch of riparian semi-evergreen forest. The path to the waterfall rose in some places above the stream and descended in some so one walked a few inches above the flowing water.

Through gaps in the canopy, one glimpsed the cloud-shrouded domes of the surrounding hills covered with grass and pincushioned with cycads. The national park boasted several botanical endemics like the ancient gymnosperm known as Beddome's Cycad, restricted to the Seshachalam Hills around Tirumala. It is eponymized after Colonel Richard Henry Beddome (1830–1911), a British officer and keen naturalist and botanist, who was the chief conservator of the erstwhile Madras Presidency.[31] *How many such gems exist along the Eastern Ghats*, I wondered, *like this little valley, with secrets yet to be unravelled by science?*

Birds are scarce in such forested regions till you hit a jackpot in the form of a mixed feeding party. Then, suddenly, you are awash in avians. On flowers, on bark, on slender outer branches of trees, in the middle of canopies, amidst dead leaves on the ground and in the air! Their musical calls blot out the din of the cicadas, and the joy of a birding bonanza breaks across your face in a smile.

Purple-rumped Sunbirds, Gold-fronted Chloropsises, Common Woodshrikes, Large Cuckoo-Shrikes, White-eyes, Pigmy Woodpeckers, White-browed, Yellow-throated and Red-whiskered Bulbuls, Quakers, Rufous-bellied, Scimitar and Spotted Babblers, White-spotted Fantails, Tickell's, Brown-breasted and Paradise Flycatchers, Shamas, White-throated Ground Thrushes, Black-headed Orioles – all vied for your attention. It was sheer bliss to follow a pandemonium of angry birds with your binoculars and focus at eye level on a Barred Jungle Owlet that stared back unblinking. Or to interrupt a Red Spurfowl in its climb uphill, and it took off in startled annoyance a few feet ahead of you. When you stopped by some bushes to listen to the song of shallow water rippling over pebbles and a Scimitar Babbler cocked its comical eye at you. When a soiree of Paradise Flycatchers pirouetted like little ballerinas in the canopy above. When you spotted a flock of Pompadour Pigeons and realized in a heart-stopping instant that they may be your pot of gold! The presence of Pompadours here increases their range and throws biogeography out of sync, or reinforces it, depending on which side of the fence one sits! They may exist in more such relictual pockets across the Eastern Ghats for all we know.

Further ahead, another surprise awaited us – the Flying Lizard, a Western Ghat species! One attracted our attention as it glided from a tree trunk to another, assisted by its orange-coloured patagial membrane. On landing, it disappeared as if by magic. Once the membrane was tucked into place, its skin merged with the bark in perfect camouflage. It was picked out again with the help of binoculars. Suddenly it displayed, and a flap of bright yellow skin stood erect from its throat!

The lizard stood parallel to the trunk, facing upwards, the upper part of its body raised on its forelegs, its throat stretched to the maximum. *Flash!* What a bright yellow flag! Again and again, it flaunted that secret hue. We realized that other lizards were around, and a patient search revealed four or five more on the trunks of surrounding trees. Was this also just another case of fitting a piece in the larger jigsaw of biogeography? Perhaps, but it was also a case of too few nature enthusiasts in this country who tramped through the 'woods' with open and inquisitive minds.

The greatest charm and challenge of natural history is that it draws us into realms outside those created by our kind. The realization that fascinating worlds thrum beyond the margins of our temporal and spiritual spheres of influence – with forms of life that are not controlled by *us* but are animated instead by destinies beyond our comprehension – holds many in awe. Places like Talakona may bestrew our land, but it is up to us to unlock their mysteries.

29

Reflections on a Short Trek to Tala Káveri

The hamlet of Bhagamandala (880 metres MSL) in Coorg is beginning to stir when we reach it at 7.45 a.m. on the last day of the millennium. We had left the Peremboocolly Estate (coffee, cardamom, pepper) near Chettalli (1,390 metres MSL), where we were staying, while it was still dark at 6 a.m. The drive via Madikeri (1,170 metres MSL) is bracingly cold and uneventful except for some spectacular views of the fog-filled stillness of luminescent valleys lit by the spontaneity of a lightening dawn. Pausing to admire one such panorama, we see humps of hills porpoising away into the distance in fading hues of blue as they merged with a silvery horizon. The air we inhale is as fresh as a cucumber and just as refreshing.

Entering Bhagamandala, we spot pilgrims returning from a dip in the sangam of the Kaveri and the first of her tributaries, the Kannike. 'Yes, there is a shortcut to Tala Káveri. You can reach it by walking there,' informs a gentleman who is wrapped in a soaking white cotton dhoti, dripping, and shivering in the early morning zephyr or perhaps with the religious fervour of his cleansing! It is believed that once a year, even the mother of all Indian rivers, the holy Ganga, travels through subterranean channels, to bathe in the Kaveri and purge herself of the detritus of sin left behind by an overload of repentant sinners. Such is the purity of this Dakshin Ganga (the southern Ganges).

A helpful police official directs us to park the vehicle in the compound of the police station. I suppose that is the safest place he can think of when we tell him our plan, promising to come for the vehicle after five to six hours! While Amitabh, my brother-in-law, drives away, my daughter Prakriti and I ready our packs, stamping our feet in the cold, watching our breath hang in visible clouds as though we're exhaling steam. Soon, Amitabh joins us, and we start off down an inconspicuous path, leading away from the open area, used by buses to make U-turns, past a school and some government offices. Here, we walk under the large-canopied trees growing on either side of the path. A Black-headed Oriole flits from one to the other, calling as it disappears inside the foliage. Excitedly, we peer into the greenery, but the golden bird has vanished.

A narrow cultivated valley lies ahead, across which begins the hill we've come to climb. The entire golden yellow field is covered with stacks of dry brown crop, neatly arranged in rectangular blocks. The rising sun lights up the entire valley like a richly textured, warm quilt. Little and Median Egrets and a Paddy Bird (in breeding plumage) peck about almost desultorily in the fallow field. A White-breasted Kingfisher glowers down its scarlet beak from an overhead wire, concentrating on the field below. In the distance, Red-rumped and Common Swallows appear like a musical score on an azure sheet as they perch in rows on stretches of overhead wires. A few glossy black Jungle Crows mope around for they still haven't found a victim to chivvy. Some confident Jungle Mynas strut and swagger in a part of the field, as though optimism at the break of day is an auspicious sign! It is at least contagious, for we stride across the fields with alacrity, giddy with our lungs full of invigorating oxygen and eager for some *good* birding.

Our destination, Tala Káveri, lies a short 5 kilometres away (eight by the metalled road), and is not the sort of place reached after hiking a couple of days through uninhabited country, where even the sound of a distant engine's roar does not intrude. But we are promised a trek punctuated with intervals of solitude and top-class, deeply satisfying birding! Since there cannot be a better way of ending one millennium and ringing in another, we jump at the opportunity.

Tala Káveri is perched atop Brahmagiri (1,360 metres) and is the source of the mighty Kaveri river. Here, ancient temples enshrine the mother goddess of the martial race of Kodavas, the residents of Kodagu (Coorg). Within the temple complex, Gundige, the sacred spring, flows into a small pond and disappears underground, emerging a kilometre downhill. On a predetermined day in the Hindu month of Kartika (November–December), Gundige makes the pond overflow in a phenomenon called Tirthodbhava (the overflowing). This is one of the most auspicious days in the lives of Kodava pilgrims, who arrive in large numbers to worship the Goddess Kaveri and pay homage to those that have departed this Earth in the past year, taking some of the holy water home. Newly-wed Kodavas sanctify their matrimony by taking a dip in the pond. In fact, along its entire 760-kilometre course, the Kaveri is considered holy ground by devout Hindus.

As we begin our trek, we find ourselves climbing a devastated slope. Habitat decay due to overuse of resources radiates, like concentric ripples on the surface of a pond, around all human settlements, and Bhagamandala is no exception. On our left, shrubs and trees have been thoughtlessly hacked for firewood. The debris of this carnage lies around on the much-used path as mute evidence. On our right is a private estate enclosed by barbed wire, protecting young coffee bushes. I wonder how long nature will allow our myopia to ravage its finiteness.

Taken aback by this unexpected scene, we grit our teeth in helplessness and hope that the vegetation ahead is intact. Pausing to survey the damage, I spy a movement in the tangle of the woody undergrowth. Black-capped Babbler! The adaptability of birds is amazing! It seems that we've stumbled upon a small, mixed feeding party. Four Black-capped Babblers rocket exasperatingly from twig to twig, denying me a clear view that is balm to a birder's conscience, till one pauses, bewildered at this contorted *thing* with enormous eyes! In that moment, I have my bird! A few White-eyes, emitting their tanpura-esque background drone, and a male Monarch Flycatcher, in his brilliant highlighter blue, are fellow hunters. It's amazing how such dazzling birds as the Monarch do not stand out in their surroundings like neon signs do in urban malls. There are others too, hunting behind the bush. A Tailor Bird *towhit*s, followed by the mysterious calls of unfamiliar birds. My peninsular birder's ear cannot decipher them. But this sense of mystery is essential to the enjoyment of the wilderness. What is there to relish in life if all is known and catalogued and the uniquely human sense of wonder lost? A lone Brown Shrike observes the scene, perched at the tip of a twig, so still in its vigil that the two seem to merge. Ahead, mercifully, the greens begin to dominate the browns.

If the outline map of Kodagu district were compared to the profile of a north-facing Nandi, Shiva's bull, the Kaveri would be the garland draped around its neck. Draining the leeward side of the Western Ghats, that stupendous storehouse of biodiversity, it flows in a southeastern direction, attracting innumerable tributaries, before it enters Mysore district. In Kodagu, the Kaveri's flow is tumultuous as it tumbles over a tortuously rocky course through narrow valleys with verdant, high banks, easing up a bit in the flatter central and eastern parts of the district, before exiting into Mysore.

Larger tributaries like the Kabini join it there, where it broadens to an average width of 300-odd metres. Its waters, used for irrigation for aeons, were first harnessed by the British in 1902 at Sivasamudram to produce 4,000 horsepower of electricity for the city of Bangalore 92 kilometres away. At that time, the transmission lines that were strung up were the longest in the world! Its waters and islands in midstream have played a part not only in Indian mythology, but also in the changing face of Indian history and culture. Srirangapatnam was the lair of the Tiger of Mysore, Tipu Sultan. Sivasamudram in Mysore and Srirangam, further down in Tiruchirappalli district of Tamil Nadu, continue to vie with each other for greater religious sanctity.

The sound of avian arguments floats in from further up the path. Walking past a screen of vegetation we see a tree with a profusion of small blood-red flowers. It is perhaps a species of *Erythrina*. The entire feathered neighbourhood seems to have descended on its branches for breakfast. Prominent in both number and belligerence are Black Bulbuls. They seem to spend more time squabbling and chasing each other than feeding. A Crimson-throated Barbet raps out his monologue from the topmost branch. Four Black-crested Bulbuls with telltale ruby spots on their throats, as though stained in the act of stealing nectar from these blossoms, guzzle nectar in the higher branches.

An hour has passed, and this seems a good spot for a bite and a steaming cup of cheer. Sandwiches, fingers of tangy, sweet oranges, wafers, and hot tea and this amazing drama of nature – better birding is hard to come by!

Suddenly, we hear rapidly approaching shrill whistles from beyond the treeline. Four male Orange Minivets burst into view overhead, pursuing a single yellow female. They seem inflamed with passion while she seems drained of all blood! There is a temporary respite for all when they too settle into the resplendent tree. Two Bronzed Drongos traipse after aerial insects from the branches, and a dashing Black-headed Cuckoo-Shrike arrives.

The blossom-studded tree stands in a private estate and may have survived the axe for precisely that reason. But here is an example of its role, its vital nuts-and-bolts utility as an ecological fulcrum in the wilderness. We savour the interdependence that hums here, the commingling of life. In the simple act of flowering – a manifestation of its sexual effusion suffused with colour, fragrance, and plenitude – the tree procreates with the unwitting help of other, altogether different life forms including insects, birds, and mammals! Unaware of their bizarre dual role in this little skit, except that of the gourmand, they descend to the orgy. From flower to flower and from tree to tree, they hop, clandestine pollinators and dispersal agents. It's sex for some and food for others. Nature works in intricate ways, and even cyberspace pales when you begin to think of the interconnectedness of a hundred million species!

With orange-scented fingertips, we brush at crumbs, rinse cups, and rise to proceed. The path begins to climb, and the estate gives way to woods that slope away from us, providing good visibility into the canopy. The hill rises on our left, also shrouded with vegetation, and ferns decorate its dew-drenched shoulder. The path itself seems infrequently used and is generously tenanted with a variety of mosses and lichens. A gentle silence breathes through the world, reflected in the wholesome satiety of the moist leaves carpeting the forest floor in the company of duetting Indian Scimitar Babblers, and in the convoy of black ants marching purposefully across the path. Conversation ceases as we realize the value and effectiveness of our other senses.

Coming around a bend, I see a Brown-breasted Flycatcher perched low in the sun-dappled undergrowth. *Muscicapa muttui!* Named after a favourite cook, Muthu, by the British diplomat and ornithologist Edgar Leopold Layard while he was based in Sri Lanka, the Brown-breasted Flycatcher leads an enigmatic, seemingly furtive existence in its peregrinations across the Indian subcontinent from the northeastern tip to its wintering quarters in the Western Ghats and Sri

Lanka. Dull brown, except for a glistening white throat, a pale lower mandible and large eyes like moist, dark brown papaya seeds, it seems uninteresting after the panorama of colour we've just left behind. It is confiding. It is quiet. It hunts flying insects with deadly accuracy, flitting after them in the shadowy region below the canopy like a well-oiled avian missile. The mystery that shrouds its life, its *ecology*, its mysterious wanderings through our country, its stubborn insistence on inhabiting patches and corridors of evergreen and moist deciduous forests at just *that* altitude knock me on to the floor. I am non-existent in its world except as the perpetrator and destroyer of its habitat. I wonder whether this exasperatingly beautiful, intricately mysterious world of ours will ever forgive humanity's hard footfall upon its crust!

Amitabh has gone ahead after an orange-skirted apparition. The Malabar Trogon has him enthralled. Neither of us has seen one this close or so clearly. Speechlessly, we gape at its exquisite symmetry. At its perfectly square-tipped tail! At the drape and finery of its plumage. At its uncanny ability to juxtapose a bit of greenery between us. What wouldn't I give to metamorphose into a branch of that leafy hillside and be blessed by the touch of a trogon!

Prakriti, tiring of the climb and impatient to end it, points out a Greater Racket-tailed Drongo. Suddenly, we are surrounded by birds. Sunbirds, minivets, flycatchers, Yellow-browed Bulbuls, and a White-throated Ground Thrush. This is another mixed hunting party that moves through the forest in an incredible show of cooperation, understanding, and non-violence. Its members cannot comprehend such human values but have nonetheless been selected by evolution to enact them! Each species exploits a unique niche for its food. There is no transgression, no trespassing, no bad blood.

The Kaveri is home to the mahseer, a cyprinoid fish with such spunk that sport fishing aficionados make a beeline for the river's banks. Anglers know the stretches of the river, sprinkled with boulders that bring the water to a boil, interspersed with deep, placid pools, where mahseers lurk. This magnificently multi-hued, large-scaled, fighting fish has a wide-ranging, omnivorous diet. It fattens up on flowers fallen from trees along the river, aquatic plants, weeds, seeds, insects, grasshoppers, earthworms, molluscs, crabs, smaller fish, water snails, shrimps, etc. It grazes in every stratum of the river, along its pebbly bed, along its coursing waters and on its surface. Fish weighing up to 45 kg were often caught sixty to seventy years ago, 5-foot monsters with powerful muscular shoulders and 38-inch girths! The record-breaking fish was hooked by G.P. Sanderson[32] and was estimated to weigh 70 kg! Officially, J. deWet Van Ingen[33] landed a grandaddy of a southern Indian mahseer in 1946 from the Kabini river; an enormous fish that was 5 feet 6 inches long with a 41-inch girth and a 10-inch maw, it weighed 58 kg!

Nowadays, fish of these sizes are the stuff of dreams. Twenty- to forty-pounders swim the waters today, mere shadows of their massive ancestors. Nirad Muthanna of Bengaluru holds the record for an Indian angler, a massive 44.9 kg beast, landed along the Kabini after an hour-long battle in September 1999. But life is not easy for the mahseer any longer. Large fish survive only where commercial angling is allowed. Elsewhere, they are not allowed to vegetate at leisure in the fertile waters of the Kaveri and its tributaries. Not allowed to hang in the swift current as though suspended by invisible threads, waiting to suck up crabs from the riverbed. Not allowed to feel the river run through them when their site fidelity propels them upstream to spawn in slow-flowing stretches of clear water. The despicable use of dynamite has devastated vast portions of the river and the life that it supports, and dams prevent the mahseer from swimming upriver to their breeding grounds.

Near the top of our climb, we emerge from the tree cover to see another estate on our right. This one has very few trees, and the plains of Kodagu stretch to the horizon. On a lone, pepper-draped, densely foliaged tree, as though clinging to the last vestiges of its disappearing world, is a White-bellied Treepie. It hunts caterpillars and swallows them with relish, wiping its beak on the branch beside it, like a truant child using a pristine tablecloth! Then, it preens its feathers. I have never witnessed a better distribution of white, black, grey, and brown on such a small canvas! I would love to loiter and soak up more 'atmosphere', as I am sure would Amitabh, but there are promises to keep.

On the edge of the hill's shoulder, above and behind us, is a tree that's taller than its neighbours. Five Chestnut-headed Bee-eaters sally-hunt insects from its branches. After a concentrated chase, they return in a satisfied, sweeping arc to their perches and swallow their snack. Again and again, they perform the same feat. In this magnificent amphitheatre, with its azure backdrop, its chequered earthy surroundings lit by the oblique rays of a rising sun, their action is so graceful in its simplicity, so unattainable in its perfection and alas, so threatened by our callousness that a mixture of emotions – rage, helplessness, futility – jam in my throat. From a branch of the same tree, cantilevered over the valley, a Golden-backed Woodpecker hangs upside down and knocks at the wood.

Emerging from these verdure vaults on to the main road, we've hardly walked a few steps when a Black Eagle, skimming the treetops, slips over us in a shallow, hesitant glide, disappearing over the treeline as quickly as it had appeared. Its brooding, dark plumage displays the bulging, reined-in power of a raptor. Its diagnostic yellow legs seem sinister in their contrast to its body.

We can see the temple complex at Tala Káveri about a kilometre away. It is noon and we are hungry. Perching on the parapet of the road, facing another valley, we pass around soft, juicy apples, bone-dry biscuits and a steaming cuppa. A Kestrel wafts delicately overhead. With effortless ease, she slips and slides on the eddies of thermal currents, surveying the slopes of the valley for victuals. At the suspicion of one, she hovers mid-flight, as though pinned against the blue sky. Dissatisfied, she planes away to another part of her beat.

Another hoverer, the exquisite, ruby-eyed, Black-shouldered Kite plays the thermal currents produced by the gradual heating of the bald hilltops. In the southern part of the Western Ghats, forests give way to grasslands after a certain elevation. But here, I suspect the denudation of these precincts by human hand, not biogeography. The last bird we see on this trek, as we descend from the temples, is a Booted Eagle, as it plummets down into the valley on bowed wings, rapidly disappearing from view.

It would be simplistic to say that the Kaveri exists because of creatures like the White-bellied Treepie and the Brown-breasted Flycatcher. But they wouldn't survive in another environment without these forests that blanket the Western Ghats. And where would the river be without these trees? And us? Where would we be, our history, our culture, the very fabric of our civilization, without the river? The presence of these birds, as indeed of all other creatures in their appropriate habitats, indicates the health of our surroundings. If we view them as traffic lights on the superhighway of our uncontrolled development, we might avoid accidents. In their happiness is our happiness. In their survival, ours. Would you then rather say that we exist because of the flycatcher?

30

The Allure of Shamirpet

For almost two decades I have regularly visited Shamirpet, a small wetland near Hyderabad. To many it may seem the same as any of the thousands of similar wetlands across agrarian peninsular India: a man-made bund thrown across a depression in the land to collect rainwater, with agriculture on the margins and below the artificial obstruction. Shamirpet's gently undulating countryside is strewn with extraordinary rock formations: weather-worn granite, in myriad shapes and sizes, piled haphazardly on top of each other, seemingly defying gravity. Low-lying areas, easily inundated through channels, are cultivated. Buffalos graze the uncultivated 'waste' areas, large swathes of which are flooded during overenthusiastic monsoons, spawning spontaneous, transient plant and animal life. Ignoring the human perception of land use or disuse, nature populates every available niche. I have spent several hours here, watching birds, soaking in the sense of the place.

It was here that I once saw a Peregrine Falcon perched bolt upright on a rock in the middle of the water, its chest glistening white in the morning sun, eyeing the waterfowl as they swam all around. Another time, a tentative Indian Courser stepped gingerly across some fallow fields on the opposite shore, giving me delightful views through the scope.

On a typical day, scores of Cliff Swallows dot the sky in frenzied pursuit of nutritious specks and then settle on wires, only to scatter at the least provocation. Graceful Small Pratincoles perch on a large, bare rock island in the water, flying hither and thither, flashing their milky-white bellies as they twist and turn in mid-flight. On a chilly winter morning, the stillness of the water echoes the fluty whistle of a Greenshank or reflects a Eurasian Wigeon's cry. On land, the handsome Blue Rock Thrush stands silhouetted on the boulders. As the sun rises higher, lark song permeates the air.

The 'wasteland' is home to at least six species. The Red-winged Bushlark and Jerdon's Bushlark shoot up into the sky and pour forth their melodies from whirred, hovering wings before parachuting down to a rock or into the grass stubble. The dapper Ashy-crowned Sparrow-Lark males court diminutive females with impressive roller coaster flights and drawn-out haunting whistles. Rufous-tailed Finch-Larks rise like sods of earth into the sky, uttering their characteristic calls, flying from one place to another. The song of the Eastern Skylark washes over those who have the ears for it, in cascades of melody. The rufous-coloured peninsular endemic, Sykes's Crested Lark moves around in exaltations of three or more, uttering its tremulous *trew-trew* call in flight. It appears suddenly out of nowhere, alighting nearby and instantly becomes engrossed in its search for food as though it was always there. What a wonderfully captivating place! For me, it is my own Important Bird Area, my personal IBA – at least in my soul if not on yellowing government records. I am sure that each of you has such a place that you visit repeatedly for what it contains, what it gives you and what you can bring back to enrich your daily lives.

Shamirpet, as I knew it, no longer exists. Development and tourism have caught up with it and large colonies have blossomed over a levelled landscape. The beautiful rock formations have disappeared under the hammer and chisel, and the water is inexorably drying up, a victim of failed monsoons and perhaps the base human greed for land. This situation is the same throughout the country, especially around expanding urban areas.

<h1 style="text-align:center">31</h1>

Birding at Anantgiri, Vikarabad

A faint *scree*, chalk-on-slate, sound comes from my right as I descend into the valley. Repeated, it piques my curiosity, bouncing an echo from a foggy memory. It is uncanny how our recumbent sensory memorabilia hibernate, coiled and forgotten in the recesses of the mind, springing forth, afresh, at the slightest provocation.

That whispered gossamer note draws me towards it. I climb a few rough-hewn steps in the rocky side of the valley, deliberately, carefully, lest its source fly away. A faint movement, a giveaway white spot on a dark rock background arrests my searching eye. Stealthily raising my binoculars, I focus on a glowing male Blue-headed Rock Thrush.

Standing erect, with neck outstretched and beak partially open, it is the picture of concentration as it creaks out that rusty-hinged strain. About the size of a Black-headed Myna, its plumage is broken up by an amazing mixture of cobalt blue, black, chestnut, and snow white, with its eyes covered by a bandit's black mask. Its mien, however, belies the swagger of its flamboyant feathers; it is hesitant and uneasy on the darkly pitted lava, and the confidence of its cut-up clothing crumbles as I approach. In a jiffy, it is off, swirling its torn cape into flight, flashing glistening white spots that mirror its irritation.

It remains in the area, despite me perching on a convenient rock, taking notes, observing other goings-on under these ancient banyan trees, rooted in aeon-defying stone carpeted with their yellowed, rustling dry leaves. I spy it often in its furtive, tiptoed flight, bluffing my existence while hunting around the margins of my vision, disappearing exasperatingly behind an obstructing branch, as though the act of shifting my head by a hair's width, for a better view, tugs it miraculously behind a screen! Who says the art of birding is a breeze? One way of succeeding at it plays upon patience: enough to earn the trust of one's target and sufficient to fool it into considering you a part of the landscape. One must sharpen this tool on the whetstone of practice to savour its fruit.

Two relatives of this elegantly feathered elfin also share these spaces. They are heftier, the size of Jungle Babblers, but more suave, every feather in place, always at their best behaviour. Their grey-blue mantle is offset by a sumptuous wash of ochre, their cheeks tear-stained white and brown, marking them for life with a palpable sadness.

It seems a busy time for them – the hour when they worm an ocean of dry leaves – and they are in no mood to let me distract them. Unlike their timorous mountain-dwelling cousin, the Blue-headed Rock-thrush, these Orange-headed or White-throated Thrushes are bolder, and so they stay. I am spellbound by their magical existence and gutted by the visceral realization that I am not even a part of their world.

They flick aside leaves with stout beaks and peer into crevices and shadows with moist black eyes. They parse every extraordinary movement tirelessly, ever vigilant, for invariably, such motion means flight or food. Today, they are determined wormers, working the leaf litter methodically, purposefully hopping on to the rough stone platform, never far from each other. I wonder about this till I see, just once, one of the pair shiver its wings in the giveaway gesture of a hungry juvenile. There is no loud fussing, like some birds I know, but neither is there any delay from the other thrush. Instantly, it is beside the now-revealed youngster, with a satiating morsel. When I eyeball this domestic scene through my binoculars, my ears filter a ratchety scratching buzz from the background noise, which is uttered in a ghost whisper by the young worm hunter.

The thing that strikes me whilst I watch these two is the method of their terrestrial locomotion. A few Jungle Babblers nearby show me how the thrushes differ in the way they hop. The babblers bounce along the ground, tackling undulations and obstructions, without hesitation or extra effort. Their legs are placed higher up their bellies than those of the ground thrushes, and I think therein lies the difference. In comparison, thrushes are lumberous hoppers. They lean forward, shifting their centre of gravity, creating an impetus, while simultaneously pushing off with their feet and launching into a laboured series of bounds from one point to another.

The thrush-mimicking babblers seem to absorb their exuberant energy directly from the earth. A ruffian flock of four rummages the leaf-layered earth pelt under the figgery, ping-ponging across its floor with the belligerence of self-appointed bouncers. They exude boundless energy as they graze the detritus-strewn land. No tindery leaf is left unturned; no crevice of bark or stonework escapes their inquisitive eye; no suspicious morsel is left uninspected. Their constant guttural banter, variable to the extreme in pitch and intensity, forms an impenetrable cloud of sound above their gastronomic safari, but their demeanour remains blasé. Propelled on springy legs, they bounce with surprising swiftness towards some unfortunate arthropod whose frantic attempt at escape, burying itself into the decomposing litter, is preordained to fail. The sweet-voiced thrush is no less a ruthless killer, nor the belligerent babbler, than a stooping peregrine flung like an anchor across the firmament to hook itself into one isolated member of a desperate flock of rock doves. Human sympathy is biased towards spilt gore and guilty of dismissing crushed chitin without remorse for the simple reason that we cannot relate to it emotionally.

Babblers banish boredom. Between bouts of babbling and squabbling, they bounce, belabour, bludgeon, bulldoze, berate, bicker, and blaspheme. They are members of the avian paparazzi, the brat pack that goes berserk at the sight of a forest-dwelling celebrity, be it a Shikra espied, an owlet discovered or a vine snake cornered. The flood of invective that they pour at the poor creature, fluttering, fluffing, switching this way and that while feverishly gripping the branch with their feet, succeeds in summoning more of their bothersome ilk to bolster their annoyed lamentations to such a pitch that the hapless victim is forced to flee, often trailing a stream of its tormentors, desperately trying to outdo each other in bravado.

Babblers are also the contact callers of a forest. They fill the pockets of stillness and quiet that suddenly envelop me in a glade with an apologetic grunt uttered intermittently, keeping fresh the breath of life that peoples the kingdom of trees. This is the way a sisterhood keeps in touch as its members glean the foliage. Hesitating on the ground before fluttering up into the lowest branch of a tree, they work their way upwards, branch by branch, leaf by leaf, till they have scrutinized the entire canopy. Having reached the top, there is a quick launch on feebly fluttering wings before the group drops, one after another, in an untidy heap at the foot of another unforaged tree. Their extreme indecision at this moment, fuelled by mercurial, enormously opinionated egos, erupts into a free-for-all of such vituperation that one would think it is a fight to the finish. They pile upon each other right there, on the forest floor, visible to all and sundry, with nothing but murderous intent in their hearts. And yet, the very next instant, they begin to preen one another, the best of babbler buddies!

Another avian marvel perambulates through the debris of shed leaves, first noticed when I am taken aback by the apparent self-propulsion of a fallen leaf but dismissed it for the wind. But it keeps happening, and when I finally focus on a moving leaf, it is transformed into a delectable Olive-backed Pipit! Three strolls among the disintegrating leaves, hidden so completely that if I look away, I lose them. They are feeding with an obvious single-minded zeal, tucking in for the long and arduous continent-spanning journey back to their breeding grounds in the Himalayas. If the globetrotters in front of me are *Anthus hodgsoni yunnanensis*, they will travel much further north, to the taiga belt. Alarmed by a Shikra's mock sortie, one bird flies up into a lower branch of the ficus, perching bolt upright for a few moments, its bold tiger-striped chest in spectacular display, pumping its tail up and down, bolstering its quaking heart. It saunters along the branch, till fear abates, and descends to feed again. Tree pipits are the silent waifs of our forests, for even when feeding in each other's proximity, they seem aloof, impersonal, introverted. These individuals in front of me, in obliterative fatigues, look rotund with accumulated fat, essential fuel for their mind-boggling peregrinations.

Looking straight up at the fecund ficus, into its filigreed lacework of a million leaves and tiny orange figs glowing in their ripeness like thin-skinned jack-o'-lanterns, I enter an uncanny planet, humming with an urgent vibrancy, peopled by a myriad world of feathered bipeds, all so utterly and completely unaware of my own living world intent upon general annihilation, that I am sucked up into their hubbub, suddenly becoming aware of the microcosmic immensity of this scene as though I were in a vision. In an unbelievable momentary epiphany, I merge with the avian metropolis above. And then, the spell breaks. A gargantuan archaeopteryx slides into my prism-enhanced field of vision. Perceptibly darkening the trillion sun specks bouncing in the foliage, it slips through the upper canopy, rending the verdure fabric of an idyllic morning with a blood-curdling shriek that momentarily freezes all feathered frolic. The Grey Hornbill resembles a giant aircraft so much, and the birds below it the gawking crowd at a busy airport, that I involuntary sink on to a rock for a moment – the world encapsulated by that fig is no less a melting pot of avian diversity than the airports of humans.

The dazzlingly bright morning starts to warm as the rising sun filters into the valley. I wander on, enchanted by sight and sound, mystified, surprised, and fascinated by turns. The valley widens ever so slightly, with space for trees and shrubbery, and tiny evanescent rills that puddle water from some recent, off-seasonal rain.

In a sun-dappled forest patch, atop a chest-high bush, perches a Brown-breasted Flycatcher. Bathed in golden light, surrounded by airy verdure, its large, pensive eyes follow every movement in the buzzing understorey, sharply watchful for flying midges. It too is piling on fat for a shorter journey than the tree pipits, and in the opposite direction too – its summers are spent on the emerald island of Sri Lanka. Its obliterative grey-brown plumage and its muted, undemonstrative ways convey a false sense of secrecy. Flycatchers of its ilk demand our focus and therein lies their appeal. They tantalize not due to any innate reclusiveness but because we are blind to the obvious. Birdwatchers also need to see!

Muscicapa muttui's aerial realm exists in the understorey of open forest, where it sorties after flying insects without any extravagant display, taciturnity being its dependable oeuvre. Once, when I press mindlessly closer to it with the characteristic intrusion of a birder, it vanishes in a smokescreen of blurred wings. Then I spot it again across the road on the tip of a declining, thin tamarind branch and raise my binoculars. For the second time that morning, incredibly, the eye streams sound into my ear. Sotto voce, *Muscicapa muttui* is muttering malignments in my direction. We see two more of them further ahead, invisible in the open.

There are erythrinas on the path, medium to tall trees, open-canopied at this time, leafless, sky revealing, but rubied with a nectary of small tiger-clawed blossoms. This little valley surely pampers its wild denizens.

It is almost midday by now and when I turn back, the frenzy of dawn tipplers has thinned to a couple of diehard Chestnut-shouldered Petronias and a few mousy-voiced White-eyes, self-consciously pushing each other towards the lambent red temptation. In the settling heat, their buzzing replaces cicada music. The petronias don't give a fig about the world. From flower to flower they move, thrusting bills deep into each cup, savouring the shiny nectar, emerging glistening beaked, dazzled momentarily by the sun, only to spot another heady tumbler. One of them, in sheer drollery, eschews the mandatory hop to another blossom's base and simply leans across space to partake off a fulsome flower, flashing its nectar-dipped beak. It is a rich sight to end a glorious morning's worth of birding.

32

Of Birds and a Menhir

I begin scratching in my notebook as soon as we're parked, the birding stuff taken out, and the cars locked. The ears direct my pencil, more than the eyes; the birds, of course, are always around, either visible, or audible.

Our large group troops behind today's guide on this birding trip to the University of Hyderabad, Professor KSR, meandering through a filigree of vegetation, towards a promised waterbody. I lean against a boulder and jot my tenth species, when a newcomer peers at the notebook and timorously inquires whether all those birds were seen today! His disbelief at my affirmative answer is expected. Most rookies do not even register birdcalls when they begin, considering them part of the background noise. He staggers away, following those who have vanished behind a screen of thorny foliage. The noisy chatter and laughter from that direction is discouraging, disturbing. Unwritten birding law frowns upon such audible excess and sudden violent movement. Breaking birding ethics is counterproductive, for the birds are scared off by loud talk. But it's okay, I guess, for this is a day of fun birding.

I too move towards the hullabaloo and arrive at an opening in the vegetation that widens into the Gundla-kunta, a small shallow pond (kunta) with floating vegetation and partly bouldery (gundu/gundam) shores. Our group stands on some flat rocks on its northeastern corner. With the sun behind us, we have perfect light for birding. There are a number of species spiritedly engaged in feeding and other activities. It is an idyllic scene – well, almost.

This little glittering wetland, so endearing to the birds that use it, is also a dumping ground for sundry items of garbage and a popular graveyard for that peculiar symbol of mindlessness, the shattered beer bottle. The place is filled with litter – empty bottles float, open-mouthed, all over the dimpled surface of the Gundla-kunta. Shards of glass glint on the shores and the rock we are gathered upon crackles with it underfoot.

This act of utterly avoidable irresponsibility shocks and angers me, for it is an indignity meted to that which I revere. This was, after all, an institution for *higher studies*. *Educated* people live and study here. Does this absolve them of civilized behaviour and its collective social responsibility? The wanton defacing of a public place needs a special germ of callousness to start it. The first bottle which is brought out into this pristine (read devoid of garbage) landscape, packed with care lest it break and leak what it holds, once emptied of its contents, becomes a burden to take back for proper disposal. Chucked or shattered, it simply catalyses similar future indiscretions. Such a sacrilegious act of despoliation soon assumes the dirty patina of a regular dump. Subsequent visitors see this 'convenience' rather than the natural beauty of the landscape beyond, which it besmirches. And before one realizes what is happening, the blight takes hold.

Should not a society that permits unrestricted access to public places also demand that such areas be left in the same, clean state in which they were found? The self-discipline that achieves this is a social prerequisite, and a really small change, in terms of the effort required for the solace of unblemished landscapes. So, what does one do to remedy the scourge? To my mind, the only solution that might last is the involvement of the perpetrators to clean their own backyard. Yes, in this case, finding the polluters might be difficult. So, involve the entire campus. Let the point of the exercise be pride in a clean campus! If this involves the teaching staff, local celebrities, environment groups, etc., so much the better! Do it every six months – I am sure it will make a huge difference, cause a marked shift in attitude, and instil a sense of belonging to the place.

While we stand for an hour watching birds, the excitement and enjoyment in the group is palpable, and why not? We spot over twenty-five species from that one spot. Besides their 'normal' activities of flying around, feeding, swimming, and perching, there is some hilarious act-like-something-one-is-not behaviour by the birds, to confuse the smug birders. There are personal highlights for me – a family of Dabchicks comprising fussy parents and three precocious black chicks cavorting in the water; both species of jacanas in different parts of the pond; and the rare Darter (two!) planing overhead in absolute abandon.

Now, there are those who will argue that this diversity illustrates the indifference of wild nature to the minor callousness of careless citizenry. But that is just a senseless, myopic argument from those who are happy to accept the visible, immediate present as the only reality. Unfettered pollution ultimately wipes out entire life cycles found in such ecosystems, big or small, gouging out the soul of a landscape, leaving behind a pauperized state of barrenness. A 'wilderness' overtaken by the detritus of an insensate community is a sad reflection of the state of a collective conscience, an objurgation of the right to call ourselves sentient beings.

Professor KSR leads us northward, circuitously, through a fallow green field of rough ankle-high grass, to the earthen bund of the Gundla-kunta. As soon as we scramble up its sloping side, an aura of an ancient presence, an antiquity, not so much geographical as civilizational, is sensed. The tamarind trees leaning away from the slope are *old*. He says there is archaeological evidence from this very campus of human settlement in the area stretching over a millennia, even of the cultivation of millets, those much-ignored cereals. It is quite likely that this very kunta irrigated nutritious crops for several generations of Dravidians. To desecrate and pollute such ancient sites of sacrosanct historical import is as sacrilegious as graffiti-scrawled monuments. What conscience permits such a strain of cultural defacement?

The atmosphere of the kunta is so different from that of the bund. Angled sunlight shines on its surface. Sounds are muted under these brooding old trees, on a ground carpeted with several seasons' leaf fall. Our wonder borders awe, and a strangely elating joy laced with deference, as though we have chanced upon a tiny gem of a marvel here, which each of us, from the glint in our eyes, seems to carry away.

The walk to the nursery tosses some more birds in our path. Jaunty-tailed Indian Robins, electrifying wagtails, a mysterious lark, a pair of amorous munias, and a fervently pleading Plaintive Cuckoo.

Tea follows, in the university's canteen, and then, near the entrance to the campus, Professor KSR leads our motorcade into an abandoned plantation of lanky-boled, thinly foliaged trees. Parking our vehicles, we walk to what turns out to be the most amazing sight of our trip, at least for me – a 4,000-year-old granite menhir, a pre-Iron Age monolith to honour the dead. It would have taken gargantuan brawn to make it stand, or remarkable ingenuity. We shoot pictures of the group around this gigantic tombstone and we disperse.

Walking back to our cars, we are appalled at the lamentable repetition of litter. All around this historical monument, this irrefutable evidence of four millennia of human life, lie the same shameful broken bottles, epitomizing the complete erosion of any respect for culture and history, of a complete lack of any sense of propriety that collective responsibility demands, of the lack of even a modicum of shame at negating personal margins of social obligation and behaviour.

I depart with mixed feelings: exaltation in the sense of perspective that history and archaeology provide; well-being from a morning spent outdoors chasing birds; and utter shame at the insensitivity of my fellow humans. But towering over all these emotions is a sense of rage and the realization that, ultimately, the education only leads a horse to water. What it does there is entirely up to it!

33

Birding with Beginners

Sometimes, it is invigorating birding with beginners. Their enthusiasm rubs off on 'old hands' like me, whose airy mantle is scrubbed off with the vim of fresh blood. This trip to K.B.R. National Park in Hyderabad has a flock of juvenile birdwatchers flexing the pinions of their new hobby. Some are photographers with forbiddingly long lenses, no doubt a handy tool even for seasoned birders. A photograph gives me the privilege of armchair identification. Others bring less extravagant equipment – their palpable enthusiasm.

Early morning perambulators goggle our motley lot in disbelief, as though we are knights of Arthur's Table. But we are used to this attention and invariably win the day by mist-netting one or more gawkers who succumb to our intriguing behaviour and unrestrained joy at sighting a bird and join us.

The first stop is near a flowering tree. It is a-whirr with wings. A family of Indian Golden Orioles; Purple Sunbirds in various stages of dressing up, resplendent shimmering males, demure hens in mute olives, and an eclipse-dishabille male in just a necktie; a ménage-a-trois of Indian Robins; a self-consciously pink-billed Tickell's Flowerpecker peering at the world through polished beady lignite eyes; a dapper cock Tickell's Flycatcher glistening in the recessed shadows, lisping a droll-whistled cadence. Also around are the commoners, chattering Red-vented- and White-browed Bulbuls, crooning Spotted Doves, with the distant clarion of a Grey Partridge cleaving all noise with its shrill urgency. Shot arrow sprays of Rose-ringed Parakeets tear the blue sky above, while giddy House Swifts chase each other's dazzling white rumps.

Where does one begin with a cheeping novice? Why, with the first bird. Sometimes, it is even more basic – how to use, no, with both eyes open (!) the shiny new binoculars. 'There, look to the right of the tree trunk, halfway up, or better, imagine the face of a giant clock in front of you, and look at three o'clock.'

To point out the pastel orange, off-white, and blue bundle that sings upon the bough, give it a name to hold on to, identify with, and watch the visible pleasure, the disbelieving gasp, is my reward. My hope is that a lifelong appetite for birding has been ignited.

Thankfully, no one argues when I point out a black male

Indian robin as one had many years ago – 'A robin has to be blue!' – ala the Robin Blue garment whitener advertised in various media. His shock at learning the truth was colossal, and for the rest of the trip he looked positively let down by the edifice the media has become.

Someone asks how I remember birdcalls. The only way I know is to follow a sound, to cut a direct path towards it, then to stop and stare at the songster, to allow its song to drench you, soak into your very skin, become a part of your existence, your charmed private landscape.

An invisible Iora whistles variations on the theme of *shaubheegi*. It blanks out my world, deafening in its lilting simplicity; its song pulses into every green fibre around me, ricochets in canopies, distils the breeze, sings of a glorious morning.

In the leafy world around me, an intense drama is enacted, covering the ambit of essential human emotions – sex, food, and territory. I am an unseeing spectator, an aural witness.

A barbwire tangle of dry, thorny branches snares an iridescent star. A cock Purple Sunbird, glistening in burnished purple, has lit upon a twig, pouring in the breeze a cascade of exuberantly tripping notes. Its hidden golden yellow epaulettes erupt into brilliantly contrasting searchlights as it sings its heart out. In the circular field of my binoculars, a blue canvas holds a jumbled thorny bush, and the mesmerizing radiance of the tiny songster. As it swivels fervently, now bowing, now stretching on tiptoe with its head thrown back, intent upon proclaiming territory, or enticing a coy hen, light dances its plumage into a coruscating iridescence.

A few 'good' birds, those that are seen well, enjoyed thoroughly, listened to carefully, and thrilled in, give me greater joy than the endless pursuit of variety, fleetingly glimpsed, abandoned, unabsorbed, and unimbued. The insatiate twitcher is ever-hungry. Here is another thought for a beginner: allow birds to come to you and savour their individuality bird by bird.

If I were to choose a seductress from the common birds to tempt someone into the birdwatcher's fold, it would be a tantalizing Golden Oriole. A resplendent male dazzles us as he feeds on the green berries of a tree. Light bounces off him with almost artificial effulgence. Unaware of his magnetic attractiveness, he paces the branches in showy preoccupation. His glistening black primaries and masquerade ball mask contrast spectacularly with his golden-yellow plumage. His strangely flesh-pink bill is an untarnished, pure appendage that is inserted delicately into nectar-sweetened flowers.

A few of us just cannot move on. We stand hypnotized by this aurum creature. What *can* one say in the presence of such beauty? We are all awed into a hushed silence of admiration. Thoughts are voiceless, internalized. All the senses combine into a single invocation, which at such times overwhelms, despite any stoicism or temporal belief, and leaves us stranded amidst the glacial progress of civilizations, the upsurge of cultures, alone, in the presence of untarnished natural sublimity. Each one of us imbibes an oriole moment in those fractional seconds in absolute solitude.

The path meanders towards a shrunken wetland that draws annual rafts of waterfowl in the winter when it is augmented with monsoonal run-off. The authorities have been moving earth in their wisdom, deepening the trough to catch and hold more water. Excavated mud is piled up as a causeway on one side. Now, the water is a shallow pool, its margins atrophied in the increasingly hot days but fringed with cooling reeds. Rounded boulders lie scattered on the shores, some under an umbrella of stunted scrub. A couple form convenient islands in the water, on which some birds rest.

The group is tired and hungry, sinking to the ground, under the leafy trees. Breakfast is passed around, throats quenched. Gradually, conversation resurfaces, and laidback birding takes hold of satiated birders.

Dabchicks float placidly upon the still water. One or two are in breeding plumage. There are a few juveniles, swimming precociously behind the adults. One, an albinistic specimen, sits on a rock in the water, foxing our rationality. Only when it slips into the liquid do we realize our folly … and it is not an albino! On the opposite shore, a grotesquely long-toed jacana flashes its cuprum wings as it scampers, as best it can, on toes that have evolved to tread on floating leaves. In the process, it uses its wings like the balancing pole of a tightrope walker. Still, it is difficult to see, and the newly minted birders learn another trick of the trade: do not move yourself but spot movement instead. The next time the lily walker flutters for balance, it is spotted. Its white eye stripe is an instant hit. Two eternally questioning Red-wattled Lapwings stride about confidently on yellow legs on the dry foreshores, yelling their lungs out. Much ado about nothing, or is it? Nature abhors wasted metabolism.

A Purple Heron sails in on cupped wings, legs dangling for a foothold, planing towards a small clump of reeds on the near shore, touching down gingerly before folding its wings over its back with visible relief. It is in brilliantly fresh plumage, ever a delight to behold. A few minutes later, its cousin, the Grey Heron appears in a slicked descent to the trembling margin. The 'essed' neck, ending at the rapier beak, is tensed up and held back, like a reined-in stallion, its stilt legs lowered to gently touch down into the pond-smelling ooze, its preferred terra firma. This individual feels the heat and wades into the water as soon as it lands, letting the stagnant marsh rise to soak its belly. It looks comical, walking thus, as though swimming! But herons hunt in different ways, and this is one, wherein prey is consciously disturbed by a wading bird and snapped up.

Our way back is partly strewn with dead leaves. The cataclysmic orgy of spring, the colossal putting forth of leaves, the landscape transforming green foliage, lies in absolute ruin all around, devastated by the stealthy turn of seasons. Stark and bare trees surround us.

'Not to worry,' I philosophize, 'this annual cycle is an engine of rebirth that is as old as the Earth itself. It sustains life.'

34

Mini Doses of Life for
Harried Souls

Areas of wilderness are at a premium in urban agglomerations, for nature is rarely allowed to exist unhampered here. It makes one accept any patch with the semblance of being a wild habitat with gratitude. Some among us harried urbanites resonate with Wordsworth's prescience, when he observed that 'the world is too much with us; late and soon, / getting and spending, we lay waste our powers: / Little we see in nature that is ours.'[34] Deep within us, we feel a primal tug that draws us to the solace that nature offers.

The exigencies of our lifestyles have taught us to celebrate wildness at the drop of a hat, in minuscule doses if need be, and in the most unconventional places!

Old walls are a favourite of mine. Layers of piled uncut stones make the best habitats.

But anything that is not cemented, or plastered with mud or lime, will also do for a home to many life forms. It's hard to believe the number of creatures that dwell in them. The sun warms up these walls, making them a favourite of the cold-blooded reptiles. Garden Lizards display their fiery red to adversaries, signalling their hold over a territory, straddled on the highest stone in the pile. And if threatened by a Shikra or a schoolboy, they scuttle into one of the innumerable little crevices between the stones.

Skinks might play among the pile of hand-hewn granite along an unused part of a compound, their smooth, dry, shiny skin creating an optical illusion of glistening moisture. Snakes and scorpions, with their fearsome and toxic armoury, live the lives of hounded recluses within deeper, darker fissures.

Spiders abound in almost every recess and cranny, safe from the winds and sweeping broom. An Indian Robin will definitely nest here if left undisturbed. Alert to intrusion, the parents carrying food to their young are duty-bound and cautious. Squirrels scamper along playing tag and other squirrel games close to trees so they can jump away to safety when a Shikra is on the prowl.

Electric poles attract Common Mynas. Those that are made of hollow pipes serve as ready-made nesting holes. Perched on top, the myna does its hilarious head-bobbing, feather-ruffling act, uttering the rusty-hinged creaking of a wind-buffeted window. Such sights are common in many parts of the city. A cousin, the Pied Myna, even dares construct large globular nests on electric poles, not only getting away with unauthorized construction but also managing to raise a family under the municipality's very nose!

Black Kites use the new tower lights that illuminate major traffic intersections in the city to build their nests. What an ideal location for them! Height, a commanding field of view, and protection from disturbance!

The gridwork of lights that illuminates the Lal Bahadur Stadium becomes inadvertent picture frames as pre-roosting kites fill them in the evening before they disperse in swirling, slow-motion tornado spirals to roost in large canopy trees.

Borrow-pits form, when chunks of earth are scooped out (burrowed) and used for construction or landfills. Invariably, these craters fill with water during the rains and become ponds or small jheels. Life creeps in and sets up a home. Cattails crowd the margins providing vital cover for timorous wildlings. Other aquatic vegetation springs up and adds its own charm. A family of Dabchicks may usurp this little soup bowl of a habitat. But its true and rightful residents are the toads and frogs. They are the visible and audible souls of such pocket marshes. The monsoon is their season, when males woo females with a lusty song. Peeps, trills, honks, and croaks resound around marshes in deep Earth-music.

A borrow-pit is a human mistake, for it helps water stagnate and breeds mosquitoes. But nature steps in with a cure – the frogs that are voracious guzzlers of mosquito larvae.

Large, open areas like the parade grounds in Hyderabad attract ground-dwelling birds. Red-wattled Lapwings strut about in undisturbed parts, rising in alarm, quizzing the guilty with their calls. Wagtails run about after small insects including mosquitoes. The population of our resident race, the Large Pied Wagtail, is bolstered tenfold or more by migrants in various hues of grey, blue, white, and black during the winter. An open space in Sanjivayya Park seems to be a favourite pad of hundreds of these visitors.

In such locations, innumerable rodents might huddle in subterranean burrows, awaiting the dark and prospects of snacking at the edible spread of human detritus. Spotted Owlets arrive with the blanketing night to keep their population in check. A late walker would hear the mouse hunter's screeches and chuckles.

Avenue trees mainly act as corridors for commuting birds. Many species visit the various parks in the city and while flying from one to another, they may stop in a tree along a road. Sometimes, these trees are able to grow undisturbed for twenty-odd years – before development catches up with them – becoming important for local fauna. Birds might begin to use them as roosts, flocking to them in vast numbers, creating a terrific din before darkness descends and they fall asleep.

The charm of natural history is that in its barest form it does not require any 'investment'. It exists. And if we create a little time and observe our surroundings, we benefit. Our lives are enriched.

As Whitman said, 'You must not know too much, or be too precise or scientific, about birds and trees and flowers and water-craft: a certain free margin, and even vagueness – perhaps ignorance, credulity – helps your enjoyment of these things ...'[35]

35

Beacons of Brilliance

There is a riot in my garden today, a riot of colour and of birds. The coral trees are a-bloom! Leafless but attired in the most radiant red blossoms, they are beacons of brilliance in the general 'brownery' of the approaching summer. Another variety, standing in a different part of the compound, has luminous white flowers that shine in the slanting rays of an early morning sun.

These trees are a veritable bonanza for insects and birds. Standing out like oases amidst the withering summery vegetation, their gaudy colours and copious nectar attract several insectivorous and nectar-feeding birds. The relationship between the tree and its visitors is obvious and mutually beneficial or symbiotic. When the birds sip nectar from its flowers, they get nourishment and, in turn, act as pollinators! It is a good idea to spend a morning near such a tree and enjoy the parade of bird life that stops by.

Some species appear out of nowhere. The Hair-crested Drongo is one such iridescent black nectar-sipper that announces its arrival with loud, jangling calls. It is a forest dweller that suffers from wanderlust and has an uncanny knack of locating flowering trees. It bullies the resident birds awhile, till its thirst is slaked, then flies on to another patch of colour. Then there are the Grey-headed Mynas, materializing out of thin air like a band of Lambadas. They flutter from blossom to blossom, dipping into the cup of amrit. Hopping along branches, they stop to peer over the side in a most comical fashion, seeming to search for something they've dropped!

Among resident birds, the most breathtakingly beautiful visitor is the male Indian Oriole. Imagine glistening golden yellow and cannon red against an azure sky! Both Purple and Purple-rumped Sunbirds sip hither and thither, unable to make up their minds amidst such bounty. Individually, their plumage will stand up to the best-dressed avian dandy.

Drongos, the resident Black and the locally migrant Grey, will not let this chance to gourmandize slip. They sip nectar less than zip after insects that descend on the blossoms like moths to a flame.

Both the crows, House and Jungle, arrive like big baddies. Their beaks are too big for the job at hand, so they whet their frustration by breaking off petals.

Three other species of mynas, the resident Indian, the Brahminy, and the noisy pink-and-black migratory

Rosy Pastor convert the tree into a noisy schoolroom. A swashbuckling, immaculately dressed Magpie-Robin stops to moisten his throat before flying to his serenading perch. A stream of assorted warblers visit and flycatchers launch after midges from the tips of branches.

A sisterhood of White-headed Babblers bustles through the branches, poking into the flowers, peering behind them, huddling briefly together on a branch, before flying away in their follow-the-leader game. Small Minivets, Cuckoo-Shrikes ... well, the list tallied by birdwatchers tops 50 species! Squirrels too love to sink their teeth into the juicy centres of the flowers and guard 'their' blossom aggressively from birds with other ideas.

For three or four days, this orgy lasts unhindered. The branches of the tree sway under the weight of those that come to feed on its effulgent benevolence. Then, its glory fades as the last few tattered flowers fall gently to the ground. It has fed many and, in so doing, ensured the propagation of its race.

36

Seeing the Spectacular in the Ordinary

'*Paan beedi cigarette! Paan beedi cigarette!*' The scratchy call of a Painted Partridge emerges loud and clear from somewhere nearby as we walk along a trail in Madhapur, an upcoming suburb of Hyderabad, adjoining the land designated for a botanical garden. Monsoonal greenery spreads all around, and the land seems to rejoice in this annual life-giving bounty, as should the people that live on it. Their joy is short-lived this year for their 'development' has backfired, and their follies have finally caught up with them. But that is not the point of this piece.

The children are walking ahead, thrilled to be outdoors. They squeal through puddles, playing those tireless games that we once played but, as adults, don't care for any longer. A smartly dressed male Ashy-crowned Finch Lark begins his display to a female sitting on the ground. He rises steeply, tipping over in a dive on half-closed wings, zooming up again, repeating his solitary Mexican wave five to six times. His whistled song jingles with exuberance. The female seems unimpressed, and he gives up and drops down beside her. The trill of a Yellow-eyed Babbler skips over the bushes, and I acknowledge its presence with surprise and wonder, not expecting it so close to a fast-changing ('developing') landscape. Then, I spy a flock of Common Babblers.

Let their name not fool you, for they are not common around our region. They are fond of this scrubby habitat and may be found on it if undisturbed. How long will this state prevail, I wonder, for the land is already dotted with granite markers and real estate developers will soon join them. A strident *teetar-teetar-teetar* rends the air as a Grey Partridge proclaims ownership rights over its territory. In the distance, we see ten to fifteen Peafowl perched on the gridwork of a high-tension pylon. They call out occasionally as they ready themselves to roost.

The life support system of this hum of activity is all around me. This ephemeral flora must complete an entire life cycle in one temporal burst of profusion. This brief span of feverish activity is the bedrock of hope upon which all other creatures of this habitat plan their own lives.

There are no imposing trees here, no glorious garden. But

life explodes here, in greater diversity than in any garden created by man. Life adapts to the climate and the soil of a region and creates systems that survive and flourish without help in cash or kind from us. It exists despite us!

Here is a ready-made botanical garden if we care to look. I see the fiery red and brilliant yellow petals of a Glory Lily nodding in the breeze from a 5-foot-tall ber shrub. An older, entirely red blossom is visible below it. As a flower ages, its yellow turns brick red as though consumed by the crimson flames of life. In a week, it is dead. The edges of its curving petals are wavy and pointed, known as *vagh nakh*, or tiger's claws in Marathi. This climber's leaf tips curl like hands around obstructions and help it stabilize and grow. The plant has developed its faculties to the maximum and makes the most of all opportunities in its blitzkrieg life.

Wild sesame plants proliferate wildly. Their trumpet-shaped purple flowers, with an extended dark purple lower lip, attract my eye. They nod 15 inches above the ground. Each flower will eventually end in a pod packed with seeds from which man first extracted oil!

Tiny solitary flowers grow atop a thin stalk. They are the creamy five-petalled Tridax. Not showy by themselves, they accessorize with butterflies, drawing the eye. When a Jezebel alights to sip nectar, the stalk bends over for the butterfly weighs too much!

Wild pink cockscombs sway gaily in the breeze. The colourful minuscule flowers form a tight papery spindle at the tip of a stem. These are often used in flower arrangements, being able to stay without water for up to two months!

There is a myriad of grasses all around. The setting sun highlights their texture, lending the landscape a gauzy, aurum hue. Grass flowers glow in the warm light as do the hairy edges of grass leaves. The plethora of vegetation within the confining habitat of a scrubland is just astounding.

There are hundreds of species here that I do not know but which would be a delight to learn about from a well-informed botanist. I have nothing against botanical gardens. Indeed, they may have their place in our search for self-education and in our endeavour to attract tourists.

But why forget what is free and part of our landscape? Why not lay a trail through such a habitat and let school children, adults, and tourists learn about the flora of the Indian landscape? One can easily design a botanical garden by spending a lot of money. The challenge, however, is in creating something spectacular out of the ordinary.

Birders

37

Humayun Abdulali: A 'Compleat' Naturalist

Among a clutch of eminent ornithologists of post-Independence India, Humayun Abdulali's path-breaking contributions displayed his active engagement with all-round ornithology and general natural history.

Humayun Abdulali (1914–2001) entered the world of natural history through oology. As a boy, his interest in egg collecting led to shikar and a lifelong passion for birds and natural history. He was so accomplished in his chosen field that the famous British naturalist Charles McCann (1899–1980) is supposed to have once commented that Humayun knew something about everything. In his passing on 3 June 2001, the birding world lost a champion ornithologist and India her most 'compleat'[36] naturalist.

Born on 19 May 1914 in Kobe, Japan, Abdulali became seriously interested in the study of birds during his college days. While studying zoology at St Xavier's College in what was then Bombay, he put together a 'creditable study collection [of birds] for his college museum from the Bombay neighbourhood'.[37] Later, this collection along with Sálim Ali's notes, formed the basis of a joint paper on 'The Birds of Bombay and Salsette', published in six parts (1936–39) in the *Journal of the Bombay Natural History Society* (*Journal*). But Abdulali had begun his tryst with the *Journal* much earlier, in 1931, when his first note, entitled 'Eleven Koel Eggs in a Crow's Nest', was published. In the following seven decades, his prolific 346 contributions to various publications covered a breathtakingly vast field of natural history – reptiles, fish, amphibians, mammals, insects, wildlife law, shikar, trade in wildlife, etc. But he wrote the most on birds, and the great majority of these papers, indeed most of them, were published in the *Journal*.

In 1968, Abdulali began to catalogue the 26,000 bird skins in the collection of the BNHS. The result was his serialized thirty-seven-part paper titled 'A Catalogue of the Birds in the Collection of the Bombay Natural History Society', published in the *Journal* from 1968 to 1996. He could not, however, complete this project due to his failing health, but he brought great zest and insight to it. This allowed him to name sixteen races of birds, of which at least nine are still in use. Contemporaries, in turn, honoured him with eponyms for two races and one species of Aves: *Pycnonotus cafer humayuni* Deignan, 1951; *Accipiter virgatus abdulalii* G.F. Mees, 1980; *Otus alius* Rasmussen, 1998 and, one frog, *Nyctibatrachus humayuni* Bhaduri & Kripalani, 1955.

But he was not just a taxonomist ensconced with museum skins. There couldn't have been a better field naturalist than Humayun. His eight trips to the Andaman and Nicobar Islands, and the resulting papers, stand as ample proof of his abilities, if proof is indeed required. His knowledge of the birds of India is amply illustrated through his full-length papers and miscellaneous notes in the *Journal*.

Abdulali was the honorary secretary of the BNHS from 1949 to 1961, a period when the society consolidated its varied inherent strengths, broke its umbilical cord with Phipson & Co., and the Prince of Wales Museum, and became an entity of national importance. We owe the present premises of the society, Hornbill House, to Abdulali's successful negotiations with the Government of India and the Prince of Wales Museum. He was also the editor of the society's *Journal* (along with Fr. Hermenegild Santapau) for volumes 57–59. Indeed, the society's welfare was very dear to Abdulali till his very last breath.

Humayun Abdulali was instrumental in drafting the Bombay Wild Animals and Wild Birds Protection Act, 1951. This was the first such legislation in India and was later adopted in almost its entirety by the Government of India for its Wildlife (Protection) Act, 1972.

The Borivali National Park (now known as Sanjay Gandhi National Park) is another of Abdulali's 'gifts' to the people of India. During his tenure as honorary secretary, the BNHS successfully represented that this area be protected by the Government of Maharashtra. An area of about 90 square kilometres was designated several years later and today serves as a green lung for the teeming metropolis of Mumbai; it is also the catchment area for the city's water supply. Later, when a national highway was planned through Borivali, Abdulali went to court against its construction. He knew the consequence of losing that suit would be the destruction of Borivali. Single-handedly, he waged a lengthy legal battle for a cause that would benefit the general public and won. Sadly, he also suffered an injury in Borivali. One day, having strayed 100 yards from the road, he and his wife were surrounded by four thugs who demanded their worldly possessions. Abdulali, of course, refused. Angered, they pelted the couple with stones, one of which struck his binoculars. Infuriated, he charged at them and got hit on his head. A friend passing by drove them to the hospital, but Abdulali was never the same again.

Abdulali showed the world that it was not necessary to be a full-time scientist to contribute to science. One could even do so while full-time in another profession. He was the quintessential amateur and was genuinely interested in sharing his love for natural history, especially with young people, and he could often be seen with them in the Borivali National Park. He encouraged people to take notes of what they saw but warned that they should verify all facts before publishing their observations as it was very difficult to rectify published matter later.

He was warm-hearted, humorous, and charming. He punctuated mealtime chatter with delectable puns and was a delightful, gracious host. Once, at lunch at his table, I was offered a whole mango for dessert. Mindful of the mess the ambrosial fruit could make in my clumsy hands, I baulked, wondering how to refuse, when Abdulali chortled and said with twinkling eyes, 'If you feel shy about soiling your hands, go through that door and tear at it in the bathroom!'

He was ready to help all those who asked him, and many current birdwatchers benefited from his knowledge. He was a 'compleat' man and a 'compleat' naturalist.

38

S is for Shikra

The title reflects Helen Macdonald's[38] visceral memoir of coming to terms with her father's death through the numbing ordeal of training a Goshawk. I've borrowed and changed her book's title here, for recalling my time with Siraj sahib has been no less wrenching. It is also fitting as he celebrated the Shikra.

In 2010, I lost two people who were very dear to me. They had studied in the same school, lived their lives in the same city, had many common friends, at least one common hobby, and both died within two months of each other in the same hospital. Both were paterfamilias of more than one 'family' – philately and ornithology. Both had a tremendous impact on my life, and both their journeys into the sun were uphill.

Siraj Taher sahib was passionate about birds. That is how we met the first time – at a meeting of the fledgling

Birdwatchers' Club of Andhra Pradesh (BCAP). The fact that he was a generation older than me was never discussed, while everything else was. The Hyderabad he grew up in, the social culture of tehzeeb (refinement) that made him who he was, has now vanished. He came to birds in the best way possible: first, pursuing them for sport, once a gentlemanly pastime, played to exacting rules; then swerving towards conservation like so many of his enlightened contemporaries. I pressed him once, to tell me why he gave up his gun. He'd shot a sambar once; a poor shot, he confessed. The injured stag blundered off into the forest. Raised in a tradition that valued a pricked conscience, he decided to follow the stricken animal and put it out of its agony. But the sambar fought the inevitable for several hours, staying either unsighted or out of range from its pursuer. When he finally caught up with it, the profusely bleeding animal was finished. The sight of the dead animal, and the realization that here was a living being that need not have died, had indeed tried to escape death, moved him immensely. With the bursting of its heart, that sambar converted its nemesis forever towards conservation.

After a couple of outings together as part of the BCAP, we began to hit it off. To this too, there was a quirky angle, which showed the apnapan (an extended kinship) of those days.

One evening, he met an old school friend of his in Riyazath Husain's iconic bookshop on Abid Road,

A.A. Husain & Co. After the pleasantries and backslapping subsided, under the convivial eye of the senior Mr Husain, Siraj sahib asked him, 'Arre, ek bachcha ata hai hamare chidiyon ke club mein. Tu jaanta kya?' (There's this boy in our birding club, do you know him?) In a city steeped in social niceties, the use of the 'tu' was reserved for those that were dear to one, unlike the rough meaning it is ascribed today.

'Apna hi bachcha hai, Siraj' (He is our kid), chuckled my father. Again, there was that tehzeeb – not the grasping 'mera bachcha' (my son), but the inclusive 'apna' (our).

At our next outing, he stumped me with a loud, 'Arre, tum bolaich nahin ki Murari ke bete hain' (Hey, you never told me you are Murari's son). Seeing me look taken aback, he chortled out the bookshop episode. After that, my birding outings received no querulous or disapproving looks at home. After all, I was going out with 'shareef log' (respectable, dignified people). Thus began a lifelong association with one of the finest people I've known. In some matters, he took me under his wing, in others we did things together.

That 'shareef' quality of the man, I think, was ultimately what drew so many people to the BCAP, which was rechristened the Birdwatchers' Society of Andhra Pradesh (BSAP) upon its registration. The hierarchy of the organization was never apparent to any of the participants. Except for Mr Pushp Kumar, the top gun in the Andhra Pradesh Forest Department, who was thus accorded a genteel deference by the elder birders, the subtle unsavouriness of implied superiority, whether of rank or knowledge, never left its residue on any of us youngsters. Everyone travelled together in the forest department's jeeps, all of us walked to where the birds were, victuals were mutually partaken of, and the day's count discussed later in a much anticipated circle of friendship. Topical and irreverent banter, and repartee, were the norm. This was an exceptional illustration of the altruistic spirit of those who planned, promoted, popularized, and participated in this pastime. To my mind, it was the single factor that contributed to the organization's successful promotion of an undoubtedly new concept of outdoor activity for the citizenry of Hyderabad. Plus, who could deny that element of thrill, shrouded in the past of fondly recalled boyhood days, of a Boy Scout's revving adrenalin, of a golden nostalgia that tugs adults to another shot at nirvana? The wilderness, and its wildlings, does such things to men.

Every specialized activity has its own terminology, and ornithology is replete with enough to confound a polyglot, its lexicon spanning classical European languages, English, Sanskrit and the immense mythologies of the world's cultures! This hurdle notwithstanding, our motley group soon cobbled together a bristling argot of nicknames, abbreviated codes, hand and eye signage and often, a whistled mimicry of surprising drollness, pulsating with distinct onomatopoeic bemusement. Newcomers stood stupefied by this bunch of loonies telling time on a tree while trying to locate a bird! Siraj sahib was privy to this notorious gang and thoroughly enjoyed the perplexity of the uninitiated. Being accepted into the group was considered a rite of passage.

His sense of fun was legendary. It often kept in good

humour a tired, flagging group returning after a day of slim pickings. A master raconteur, he spiced tea breaks with stories from days of yore, in turn becoming moist-eyed and sentimental about vignettes from *The Days of the Beloved*[39] or rasping slang Hyderabadi in his baritone.

For Siraj sahib, every birder was special and welcome in his home. I met many heavyweights of the birding world there and cannot recall a single person who was not at as much ease as they would be in their own drawing room. The city's birders walked in and out as if they were visiting their own homes. And to me, it was always a special place. I went there for all the reasons that a deepening friendship draws two people to spend time together. Our discussions ranged from ornithology to social behaviour, politics, the philosophy of morality, ethics, the nuances of Urdu shayari, the importance of art in comprehending the world, the unceasing turmoil that was the lot of our environment, the way to take society forward – several ideas sprang up spontaneously during conversation, which we worked on and made something of. But the deepest moments were the silences we lapsed into between words, ruminating, pondering, or simply savouring the thoughts that rush in at the end of a satisfying flow of words. I always took away the comforting thought that this was someone I could speak with about subjects and thoughts that shone with clarity once the polishing strokes of give and take had burnished their cantankerous edges.

The heady plan of collating our own checklist of the birds of Andhra Pradesh was cast during one such evening. Why couldn't we do it? It was the first step to more comprehensive ornithological studies, but this could be achieved by just the two of us. As often happens, the force with which an idea hits one, it also pushes out and lays bare methodologies for fructifying it in the same instant. I do not think it took us more than half an hour to decide our workflow and apportion our responsibilities. Subsequently, there must have been innumerable meetings to iron out creases, enjoy quaint discoveries in published accounts from the past, streamline every step so that we were both on the same page with regard to content, presentation, citation, taxonomy, etc. But there was never really any major change, no rollback of methods, no U-turn on our road map.

Comparatively, translating *Birds of Southern India* by Richard Grimmett and Tim Inskipp (2005) into Telugu was a minefield. Collaborations were imperative because he had only a good working knowledge of the language and could be the ideal Telugu-speaking ornithologist, but was not proficient enough to take on the entire work himself. Translating the extensive lexicon of birding into a language that seemed to have no specific, definitive words to suit scientific interpretations had to be executed by a double-nibbed pen: one that wrote with bold, hearty strokes, delineating the work; the other, with the fine point, working in all the telling details that illuminated the manuscript. I was the troubleshooter who tried to unclog detritus in the workflow to ensure smooth progress. Not least of these was dealing with the names for the colours of plumage, 'brownish yellow', 'blue green', etc. Ultimately, we did get our own Telugu colour palette! Siraj sahib worked hard and long at the manuscript, and he may have finally taken it through at least six proofs after nerve-racking meetings with the translators. The fine-pointed pens scratched away day and night, and I would notice fresh stacks of papers during our daily meetings. He had this thing for coloured inks and would mark up his sheets with differently coloured inks, each linking a unique chain of thought. He wrestled with extant Telugu bird names, which were invariably of a generic nature, and had the distinction of coining dozens of new ones. But it was mighty frustrating work, and I often did catch him in a reverie, absent-mindedly sucking at the end of a pencil. The lack of a consensus, the engine that could have driven this naming work, rankled him. Invitees never bothered to respond, except a few, for which he was so grateful. Dialect, that verbal filigree of distinctiveness within a language, a culture, tortured him till one day he realized the only way forward was to ignore it. One could not please everyone. When it was finally published in 2007, it was a work that justified his sense of achievement, in that of his team, and in his pride that the society was a co-author.

Siraj sahib, and indeed, Pushp Kumar, worked with an extraordinary zeal to promote an awareness of birds among Hyderabad's citizenry, and as a consequence, shone a light on the environment we collectively belong to. The media, both print and radio, smelled the potential of a unique angle to pitch stories, and people began to discuss birds, the urban environment, and even the activities of the BSAP! Birds have always attracted mankind, and when brought out from the realms of literature, poetry, art, or even mythology, into the ambit of casual conversation, they ushered delight, rekindled memory, and nudged people to look around and notice the natural world. The quiddity of our surroundings during our growing-up years is deeply ingrained in our bones, recumbent though it may lie as we rush about our busy days. But there comes a moment when memory stirs, awakened by a petrichor sprinkled upon our passivity by people like Siraj sahib, and a delightful facet of the natural world is miraculously visible to us. Birds can charm the most indifferent of people!

If passion fuels altruism, the result is joyous. Siraj sahib's involvement with the BSAP was such a phase of his life. Whether he was organizing a field outing; writing the next month's programme in long hand on *all* the postcards that he would mail himself at the post office; correcting proofs of the society's publications; sitting with the patience of a Buddha for various permissions from a forest official; answering mediapersons even when they came unprepared and posed prosaic queries; preparing environmental assessment reports of places he deeply cared about, like the Kolleru Lake, and sending them to the concerned authorities on behalf of the society; speaking to students; chairing symposia – it was all 'for the birds', as though he had an oath in heaven!

And when his penchant for the living bird needed a rest, out would come his philatelic collection with *birds* as its theme! He worked on it whenever he found time and even displayed it successfully (winning competitions) in several philatelic exhibitions. The forest department regularly petitioned him to display it during their annual Wildlife Week celebrations, and he would willingly oblige. After all, it presented a fantastic opportunity to convert the visiting schoolkids to a lifetime love of birds and philately!

He took great pleasure in the company of lifelong friends, in the earthy delight of seasonal fruit, in the song of garden birds, in the pressure of his granddaughter's tiny hand around his finger. In a sense, he did see the world in a grain of sand. When the end came in the January of that fateful year, he had a fitting farewell. A gentle evening descended; the annual numaish (fair) that he enjoyed with his beloved Ayesha flung high its illuminated giant wheel; Mukesh's poignant, pathos-filled '*Jeena yahan, marna yahan*' wafted from that direction; and as clods of his hometown earth sprinkled the path of his onward journey, a pair of spotted owlets, that he endearingly called *chakwa–chakwi*, exchanged yarns inside a dusty tamarind.

39

Remembering Zafar Futehally

An Indian naturalist and conservationist Zafar Futehally (1920–2013) started the *Newsletter for Birdwatchers* and edited it for forty years. He was the recipient of various awards including the Padma Shri (1971), the Order of the Golden Ark (1981), and the Karnataka Rajyotsava Award (1983). Two incidents from his life are recounted in this essay.

Zafar sahib was a diehard optimist. In 2004, after he lost control of his beloved *Newsletter for Birdwatchers*, he called me one day and said, 'All is not lost. We could start all over again.'

For the benefit of those whose interest in ornithology took flight in this millennium, the *Newsletter* was born on 31 December 1959. During an evening walk in Bandra, Sálim Ali had suggested to Zafar Futehally that the time had come to form a Birdwatchers' Field Club of India and produce a '*Newsletter for Birdwatchers*', which would consist primarily of fresh observations from the field. With characteristic commitment and zeal, Futehally plunged headlong into the project, taking on the onus of producing it, and he also became its first editor. Its annual subscription was ₹5. Gradually, its readership grew, despite the issue being slim, cyclostyled and stapled together. Zafar sahib encouraged rank amateurs to publish in it and cajoled well-known Indian and foreign ornithologist and naturalist friends to write for it. Sálim Ali was a frequent contributor to early issues. The *Newsletter* became a repository of a fascinating plethora of birding notes from across the country, and Zafar sahib's liberal editorial policy encouraged contributors to send in their notes and papers to it regularly. Many of today's senior birders cut their teeth here. Its correspondence columns were like a coffee room abuzz with bird talk.

As he got on in years, the workload must have become substantial, and people volunteered to chip in. S Sridhar of Bengaluru took on the laudable task of printing the *Newsletter* and in the September–October 1988 issue, after nearly twenty-eight years of distributing cyclostyles, Futehally's editorial triumphantly announced, 'At last in print.' Over the years, Sridhar also took to managing subscriptions, but Zafar sahib remained the editor.

In January 2004, at a gathering of its subscribers in Bengaluru, Futehally announced a change in the editors of the *Newsletter*, emphasizing that Sridhar's role as publisher was irreplaceable. He asked the gathering whether anyone disagreed with this change – none did. But when the January–February 2004 issue of the *Newsletter* was published, it was clear that the four-decade-old publication had new people at the helm, with a new editorial board instituted by its publisher, without consulting Zafar sahib. His only comment, when he had a copy in hand, was 'I have received a few shocks in my life, but the arrival of the Jan/Feb issue today has been a special one.'

Soon after, Zafar Futehally, V. Santharam, and I set up the New Ornis Foundation (NOF). Santharam had been a regular and prolific contributor to the *Newsletter* since 1978, and his detailed observations of birds and their behaviour were a treat to read. After graduating in commerce, he did a master's in ecology and a PhD on woodpeckers. He teaches at the Krishnamurti Foundation school in Rishi Valley, Chittoor.

One incident in particular, from the time we registered the trust, has stayed with me. The three of us founding trustees agreed upon a mutually convenient date for the registrations, and on D-day, 25 October 2004, the other two arrived in Hyderabad, and we went to the sub-registrar's office to register NOF's trust deed. Later, at lunch, Zafar sahib peered at us with mischievously glinting eyes and said dryly, 'You know, I nearly didn't make it today.' When pressed, he confessed that he'd gone riding a day earlier, and the horse had thrown him. He would have been eighty-four years old then. 'I was in a lot of pain, and so I quietly called my doctor and informed him of the mishap. But I told him I simply had to go to Hyderabad and entreated him to ensure I could go. I also warned him not to tell Laeeq [Mrs Futehally]. Thankfully the pain subsided, and here I am!' That was the level of his commitment and passion.

Another incident about him that comes to mind is from 1998, when Indian birdwatchers and ornithologists were meeting in Coimbatore. BirdLife International (BLI), the world's largest global partnership for nature and people, was in town to enrol one Indian organization as its international partner in India. BLI started as the International Committee for Bird Preservation (ICBP) in 1922 and became BirdLife International in 1993. Today, it has 115 partners worldwide, besides over two million birders, scientists, and local volunteers that help it study and understand every bird species in the world.

The toss-up for such a candidate was between the BNHS, and the Sálim Ali Centre for Ornithology and Natural History (SACON). Three days of back-and-forth conversations did not seem to resolve anything; talk also veered towards a new, democratic pan-Indian organization as one of the BLI's criteria for selecting a partner was that it should be autonomous and membership-driven. In the concluding meeting, the general mood was cloudy, and amateur birders were justifiably discouraged.

V.S. Vijayan, the director of SACON and our host, announced that Futehally had to catch an early flight, and wished to address the assembly for a few minutes before he left. Midway through the event, he strode into the marquee, as the seated audience watched, and was given the podium. Gentlemanly to the core, he apologized for his interruption, thanking everyone present for tolerating the intrusion. Then, in less than ten minutes, he summarized the proceedings of the past three days, explained the bottleneck we were in, and gave the audience hope that only good could come of so much genuine deliberation. He walked away to a standing ovation. In 1998, BirdLife International chose the BNHS as its India partner (the outcome of a strategic planning workshop in September that year), which agreed to form a sustainable network, linking ornithologists, birdwatchers, and organizations working for bird conservation in India called the Indian Bird Conservation Network (IBCN).

This was Zafar sahib's special gift. He cut through complex situations, and seemingly impermeable bottlenecks, to the core issue at hand. He saw things and situations plainly, stripped of all their emotional and sentimental trappings. It is not surprising, then, that he was an integral part of independent India's early conservation history, from shaping policy, steering the BNHS (as honorary secretary), editing the iconic *Journal of the Bombay Natural History Society* (1963–73), setting up World Wildlife Fund-India, starting and single-handedly running for nearly four decades the *Newsletter*, or working towards the preservation of the lakes of Bengaluru. His sole work ethic seemed to have been to get on with the job at hand and see it through to its logical end.

Zafar sahib's diplomatic skills were indeed formidable, especially when coupled with his indefatigable energy for pursuing his 'passion projects'. By using both for India's voiceless denizens, he made this a better world for all of us.

My Kind of Birding

40

Why I Birdwatch

The water floats about 400 ducks. It is still, and a dark grey-blue, under a perfectly clean January sky. Four of us are huddled behind a mound of earth on a man-made earthen bund with large-canopied tamarind trees lining the water's margin. Driving two hours from Hyderabad, the landscape pouring away from our speeding car, we have come to this little wetland to count birds. The sudden stillness of the landscape is very welcome; its grounding elemental stability and the warm sun on our backs is comforting.

As we settle into the basic mathematics of a census, the southern sky is suddenly stitched with uneven lines of ducks. Skeins approach this lucid Earth's eye, floating like motes from one end of its curving firmament to the other, turning in tight formation, tugged back by an instinctive bond with their swimming brethren, as though an elastic band were stretched to its maximum and then relaxed. Twice they circle the water at breakneck speed with a completely mad effortlessness; two peel off, making for the water. A cohesive tension in the flock breaks, and the entire lot plunge earthwards in comic pell-mell. Their sharp descent rips aloud the blue sky-cloth, and in the enveloping silence, the tearing sound of wind forced through curved feathers is an astonishing aural delight.

Is there a palpable joy in these returning anatids? Open-mouthed, I gape as one, then three more flip in their descent, turn turtle, belly up in their plummeting fall through the ether, righting themselves at the last minute to do a belly flop in the water. It's a remarkable feat of contortion. While the body rotates upwards, twisting at the neck, the head is held straight to retain orientation. Why would a bird do this? I grope for logical answers, then settle for anthropomorphism – for the sheer thrill of it. Happiness may not, after all, be an entirely human privilege.

From the moment I spot them, through their fantastic airmanship, the magical rending of sky-cloth, to the jubilant splash of sparkling water, I sit mesmerized by the beauty of the moment. It is burned indelibly on my mind's eye, etched upon my heart, and I think this is reason enough to watch birds.

The Red-crested Pochards bob like cork in the eddies their landings create. There is great excitement and agitation in the roughed-up raft of Anatidae. Drakes rear up in the water, stallions on hind legs, flailing their wings vigorously. This flotilla of punks, dressed up in luminous orange headgear and candy-pink beaks, is riveting. Gradually, peace returns to the waterscape, and ripples plane out. A jostled breeze sighs through a million tamarind leaves. Borne on the wind are the whistles of Eurasian Wigeons flaunting their sandalwood paste–smeared foreheads in the distance.

It is a wonderfully idyllic setting. In the lee of the bund stretches a square patterned paddy tapestry, embroidered with glinting still waters. Its miniature dykes are patrolled stealthily by rapier-beaked Pond Herons, its sky-reflecting water frogeye dimpled. Forlorn toddy palm clumps stand abandoned in scattered disarray across this quilted landscape.

Screaming Palm Swifts tail each other, swing around fronds, arrow from clump to clump and boomerang back on slender fluttering wings. Small Green Bee-eaters catch their prey with an audible snap of their mandibles and float down to a telegraph wire to dismember a bee of its business end, or disrobe a butterfly roughly, tearing off its wings, before tossing the naked body down their gullets. The buoyant colourful wings, plucked and discarded, bereft of direction, abandon themselves to the wayward breeze.

Bullock carts clatter past to a village at the end of the dirt road. A skeletal, barefoot toddy tapper pedals his spare bicycle towards lanky trees, in the crowns of which he has tied a neat cluster of earthen pots to collect the oozing liquid that gradually ferments into toddy. He travels light; loincloth, turban, sickle, circle of climbing rope over one shoulder to hoopla around the thinly corrugated palm, and the throat-scratching packet of rolled beedi leaves. He passes by trailing a haze of mild fermentation.

Birding takes me to such serendipitous settings where the zeal of citizen science takes a back seat, and my senses absorb an essence of landscape I otherwise rarely absorb. It creates wonderful opportunities to delve deep into myself, stilling a futile search for life's meanings in an increasingly fractious world that we are urbanizing at catastrophic speed, taking away from a generous land the solace of open spaces.

41

Birds in Our Cities and Our Lives

A common question that people often ask me is, 'Where do you go to watch birds?' Their surprise borders on disbelief when I tell them that I need not go anywhere. Within the urban sprawl of Hyderabad, at least 200 species of birds can be seen. In our parks and gardens, around our lakes and jheels, in our monuments, bungalows, and apartments, on our roads, and around our garbage dumps, in our grain and fish markets – birds abound in the city. They are a part of nature, the greatest show on Earth, and the best part is that anyone so inclined can see them. There is no

charge, no entrance fee, no seat number. But do we have the time to stand and stare?

However, with increasing urbanization and the conversion of every square inch of available land into real estate, we are destroying habitats faster than we realize. Let us look at what is happening and how it can be remedied in a little more detail.

Vegetation is vital for birds. They use shrubs, creepers, and trees in many ways. Fruit and nectar provide nutrition and energy. The limbs of plants provide perches, places for nesting and socializing, and serve as lookout posts. Dead trees are invaluable too. Woodpeckers and other hole-nesters like barbets use them to raise families. Insects that thrive on these trees are consumed by woodpeckers and other species. But we do not like dead trees in our gardens and urban parks. Dead leaves carpeting the ground provide cover for insects that birds like babblers feed upon by hopping around, flipping over leaves, searching for critters that hide underneath. If fallen leaves are left unswept and unburnt, we will enjoy the sight of a flock of babblers rustling through them, and, who knows, an Indian Pitta, also a gleaner of ground cover, might grace us with a rare darshan (appearance).

Therefore, to attract birds to our gardens or parks, it is essential that we plant trees and shrubs that are native to our country and that bear fruit and flowers palatable for birds. A list of such plants can easily be procured from the state forest departments, or even from your favourite nursery, or botanical society.

Water is another vital requirement for birds. They love

it. To attract birds to our homes, we should provide them with water. The sound of falling water draws birds like a magnet. If a shallow pool can be constructed in a garden, wonderful! If not, a flat vessel can be kept out in the garden, veranda, or on the roof – anywhere that's convenient, and accessible for birds. The water should be changed daily. Such artificial sources of water should be placed such that a bird can make a quick getaway from cats and dogs. Besides slaking their thirst, they enjoy a bath and will indulge in one daily if they have access to water. City pollution clogs and damages feathers, making them unresponsive during flight. A bird must always keep them in their prime. After a bath they spend much time preening each and every barb of their feathers, oiling them if necessary.

The protection of our lakes and jheels is our duty as they recharge underground aquifers, besides being natural drinking and bathing places for wildlife. The ponds and small waterbodies in our cities should be guarded and protected by those who live around them.

Many species of birds nest within city limits. Most of them find safe places in trees and bushes, collecting nesting material from the wild besides selecting objects that we use. Thread, bits of wire, broomstick twigs, plastic, cloth, shiny beads, cigarette butts, and in one case, gold-rimmed spectacle frames, all find their way into nests! Many hole-nesting birds suffer from want of suitable sites as trees with holes are rare. This is so because holes may form in diseased, dying or old trees that are pulled down because they look 'ugly' to the manicured mindset of urbanites. Barbets and woodpeckers excavate a hole for themselves and at times these get enlarged and are taken over by larger birds like Grey Hornbills and owls.

As I have said earlier, leaving nature to take care of itself is the best philosophy. But most of the time urbanization does not allow that. At such times, nest boxes can be made, or bought, and put up for birds. The technical details to make one are readily available online, or one may contact a local birding society. Nesting material, consisting of swept twigs, fallen hair, thread, etc., can also be left in a corner of the garden for the birds to use.

A bird table is a common feature in many western gardens. People put out food for birds, especially in winter, when the weather is harsh. This is generally not necessary in India, but if one is so inclined, kitchen scrap can be placed regularly in one place in a garden. A simple bird table can also be put up, but it should be cleaned regularly.

The Care of Injured Birds

Birds are extremely delicate creatures whose bodies have evolved for the type of life they live, and they live life at such a hectic pace that accidents are bound to happen. There are various ways in which birds might get injured. A bird might have escaped an attack by another bird or animal, but sustained injuries in the process. It may have had an accident while flying. Birds are often confused by vertical glass on buildings and dash against them in mid-flight, ending up below them either dazed, injured, or dead. They are frequent victims of the glass thread (manja) during kite-flying festivals. Nestlings may fall out of nests. Migratory birds that stop to feed in our cities before flying onwards, or those that arrive to spend winters, are tired and disoriented after unbelievably long and arduous flights. They are depleted of energy and weakened by their magnificent exertions. In a new neighbourhood, they have to fend off local bullies (for example, crows and cats), and such stresses and traumatic experiences take a toll on their health. It is best to leave them to themselves rather than try and rescue one, for your intervention might injure the bird you are trying to save.

Human sympathy towards nestlings that have fallen out of their nests, or fledglings that are out taking their first steps in the world, is universal and generally misguided. If you happen to be in such a situation, please do not pick up the bird, however good your intentions. It may be in the process of learning how to fly. Its parents could be around with food and resent your intrusion. It is best to let nature take its course. Inadvertently, in trying to be a good Samaritan, you might do more harm than good.

If you must capture a bird, let it be only one that is physically injured or visibly traumatized. Great care should be taken not to injure a bird while capturing it because any extra pressure on its body can cause injury. The wings should be held close to the body and not allowed to flail. You should yourself take care and not get injured by thrashing feet or snapping/biting/lunging beaks. The powerful beak of a parakeet or a shrike can cut open your hand to the bone. You see, it is easy to capture a bird, but not so easy to take care of it. If the bird is seriously injured – a wing or foot might be broken – rehabilitation is very difficult, and the bird generally does not live.

If a bird that has been caught is not physically injured but appears sick or shocked, you should keep it overnight in a well-ventilated box (a shoebox with breathing holes in the sides will do) in a quiet, warm, and dark corner, away from pets. Try and identify the species and give it water and proper food after it has rested for about an hour. If the bird is a seed-eater (sparrows, pigeons, munias), then a mix of birdseed will do. If it is an insectivore (myna, bulbul, etc.), then you will have to procure worms from a vendor of ornamental fish or feed it finely chopped boiled egg mixed with crumbs of biscuits or bread. Minced meat will also do.

Visibly injured birds, or those that are very sick (those who do not eat or drink and allow themselves to be caught without any resistance), should either be allowed the space and dignity that your inaction provides or be taken to a qualified veterinarian immediately. Most such cases are hopeless and may have to be euthanized by the vet. It is the most merciful thing to do.

The aim of rehabilitation should be to release the bird in the healthiest condition as soon as possible and in the most appropriate habitat for the species.

Caged Birds and the Release of Captive Birds

The Government of India banned the trade and trapping of wild birds in 1991, and any such activity is totally illegal today. And yet, the trade thrives. Birds are bought and sold for at least seven different purposes, none of which is beneficial to any bird. In fact, the rate of mortality among traded birds is shockingly high: a large number must be captured so that the few that survive can be sold. Trappers use numbingly cruel methods to subdue the wild spark of the captured bird.

We keep birds as pets. We eat them. We use them in medicine. We display them in zoos. We display their feathers on ourselves, or in our homes. We pit them one against the other and call it sport. We dupe each other by using birds as vehicles of black magic and voodoo. We release captive birds and bask in the satisfaction of a good deed done. There are many socio-cultural issues here, but I would like to touch upon two that form a large part of the bird trade in our country.

Compassion and the welfare of the birds are the last things on a bird trader's mind. Parakeets are de-beaked by rubbing the points of their beaks in gravel and sand, so that their bite does not hurt human hands. Sometimes, their lower mandible is dislocated for the same reason. The gullible customer does not know that in a day or two their dear pet will die of starvation anyway. These are just two examples of what birds suffer at the hands of men. For every bird sold, ten would have died, if not more, before they reach the market. Birds are also transported in the most despicable conditions. Many have their wing feathers pulled out, so they cannot fly. The trauma and stress that these birds undergo is unimaginable. How can a true bird lover ever be happy when they look at a caged bird?

The custom of releasing a captive bird is deep-rooted and widespread in Indian cultures. To release a captive bird, and deem it a good deed, is intrinsically wrong. It is a self-delusional act that looks at just one angle of a problem. The trader gets his business, the releaser his misplaced sanctification, and the birds keep on dying. If the proud owner of a pet has a conscience, these deaths should prick it. It is their purchase that keeps the trade alive. It's a vicious circle that can only be broken by us when we realize the futility of our deeds. To end the misery of the birds, we must stop buying them. We must learn to step out of our houses and enjoy the thrill that a bird gets from its freedom. To appreciate and understand this is the best thing we can do for the bird. For this, we must think deep and answer our conscience: Are birds the true beneficiaries of our actions?

42

Checklists: Ticking it off

It is a peculiar character of human nature which urges people to collect. There are those who collect stamps, and then those who collect fancy books, bottles or even pieces of driftwood. Indeed, though each man is an island unto himself, given a chance and the means, he will also become a museum unto himself!

There are also those who collect bird notes and lists of birds seen in various places. Bird notes, when compiled properly, can shed light on the life histories of our avian friends and on the dramas and little intrigues which unfold in their lives each day. Bird lists help us fathom the ecological viability of an area, a habitat or a geographical feature. Despite their many uses, the only problem with listers is that they seldom assemble their notes into one comprehensive list for an area – so much so that ornithologists often despair that the world is too much with listers!

A Checklist of the Birds of Andhra Pradesh began when Siraj Taher and I realized that there was no such data available for the state. Not that the avifauna hadn't been studied; but the results were scattered among various libraries and not easily accessible to birdwatchers. Dr Sálim Ali's Hyderabad State Ornithological Survey in 1931–32 covered the birds found in the erstwhile Nizam's areas. The results were published in the *Journal of the Bombay Natural History Society* more than fifty years ago and are now practically lost to present-day birding enthusiasts. The studies done by Humayun Abdulali in Visakhapatnam, and the more ambitious Vernay Scientific Survey of the Eastern Ghats suffered a similar fate. Work on the avifauna of Andhra Pradesh has continued intermittently since Sálim Ali's survey, mostly restricted to small parts of the state – the Eastern Ghats, Adilabad district, etc. But here too, most of the findings are not easily available.

Then came Sálim Ali and Dillon Ripley's ten-volume *Handbook of the Birds of India and Pakistan* (hereafter *Handbook*). It was exhaustive, monumental, prohibitively expensive and difficult to carry around in the field (and also currently out of print!). Unless, of course, ten birders go out together carrying a volume each – nine of them watching, while the tenth refers to the pertinent volume! A student of ornithology would have to dig through more than a hundred references from various journals, newsletters, and magazines before even beginning to draw any picture of the bird fauna of the state (we arrived at the figure of hundred after our searches, and there are likely to be many more that we missed). This would be a formidable task indeed, and we resolved to make it easier for him.

Comprehensive checklists are the basis for the understanding of any regional avifauna and reflect its environmental strength and vitality. They can be drawn up for areas marked by political boundaries or for geographical features of the land. Rather than assemble a list of birds of, say, the Eastern Ghats or the Coromandel Coast or a wildlife sanctuary, we decided to muster one for the political state of Andhra Pradesh. It would cover all these areas and perhaps be more useful to birdwatchers, whose perspective of a region is more often political than zoogeographic. We divided the state into three major physical regions – the Eastern Coastal Plain, the Eastern Ghats, and the Deccan Plateau. Each zone became a criterion for the distribution of bird species in the state. We worked from the *Handbook* and jotted down those species which occurred in Andhra Pradesh. Gradually, data on distribution and additional notes on behaviour accumulated, and the checklist took shape.

One problem was that while we were concerned only with Andhra Pradesh, the *Handbook* dealt with the entire subcontinent. It was only occasionally that Andhra Pradesh or a place within it was specifically mentioned. But more often than not, we were swimming in the 'entire peninsular India' or floundering 'from the Punjab down to Kerala and east to West Bengal'! Such entries required more spadework to support their inclusion in the list. An even more trying aspect of the exercise was that Andhra Pradesh was created during the reorganization of the erstwhile Madras Presidency and the states of Mysore and Hyderabad. Previous surveys had been sponsored by the respective heads of state and restricted themselves to relevant state boundaries, and we had to reorganize much of the data.

Getting more reference material together was another game altogether. We combed libraries, both public and private, for old journals of the BNHS, Zoological Survey of India, etc. We poured over the contents, the various papers, and notes, often wading through protracted controversies about the races of birds. Some we were able to resolve to our satisfaction. Those that were too involved to unravel easily were clubbed together with a query, implying that either of two given races could be found in the area.

Checklists are useful because they contain two vital pieces of information on the different species – distribution and status. Checklists from areas that are birded consistently, e.g., a city, or a district, over several years, gives some idea of the status of the birds therein. The status of birds for large areas may, however, have to be generalized, for detailed information is not available in our country. But such data, which covers pockets of an area, is useful and may become critical in bird studies like patterns of local migration. The inclusion of bird names in the vernacular dialect is useful in the field when the assistance of locals becomes essential.

There are some points which one might say are prerequisites of a good checklist. One, entries should be cross-referenced to an accepted reference work (we used the *Handbook*) to standardize information. Two, the authenticity of the records which have not been published in journals should be verified with great care. And even the printed word is not infallible: it too should be scrutinized. Three, a complete bibliography is essential. Besides imparting an aura of authenticity to the checklist, it might also turn out to be the first complete assemblage of bird literature for an area! Asking help from a senior ornithologist can resolve many a confounding issue quite easily. Indeed, listers' discussions sometimes go off at such unusual tangents that the penetrating insight of a trained mind is pure ozone!

All this work must be presented in a slim volume that is handy, fieldworthy, and affordable. In our country, it is not wrong to say that checklists will have to be constantly updated for some years. Birding in India is only just taking wing, and records will keep changing as more birders take their hobby and their notes seriously.

A checklist can replace neither the field notebook, which must be the primary source of all information for the naturalist, nor the field guide, used for on-the-spot identification. But it is a handy reference in which notes and sightings can be jotted down. It certainly helps a birdwatcher understand the birds of an area a little better.

43

Birding by Ear

Birding by ear is as enjoyable as classical birdwatching! I was introduced to this facet of the birder's art, very early in my birding journey, by friends at a World Wildlife Fund-India camp at the National Defence Academy at Khadakvasla in Pune. On most of our outings I was nonplussed at the longer bird lists that my friends read out at the end of the day. When I wondered aloud how I had not seen such and such a bird, I was let into this birder's secret. They had heard the birds which I had not seen! My endeavour since then to lend an alert ear to avian phonetics has paid rich dividends.

The trick is to follow any bird call, or for that matter, any intriguing sound in the wilderness, to its source. This is the surest way of identifying it. I remember how some of us spent half an hour under a huge tree in the Eturnagaram Wildlife Sanctuary, heads craned upwards, perplexed by a loud,

"

ringing call, repeated again and again, till someone spied the Indian Cuckoo, the 'one-more-bottle' bird in the foliage!

Birdsong is a source of delight and information. It is a fairly accurate indicator of the whereabouts and circumstances of the caller. With care and patience, one can distinguish between the alarm call of an Ashy Wren Wren-Warbler and its song. As one becomes proficient in this art, the pleasure of birdsong is felt deeply when the vista on a hot afternoon is nothing but a spread of babool and Prosopis, and the easily identifiable raucous call of Large Grey Babblers wafts towards you on the tendril of a breeze. I have delighted in the morning chorus of various places and maintained lists of bird calls as they occurred, one by one, with the approaching daylight. Many a time, fellow birders slept through it all! It becomes second nature to identify the caller in the mind, while within the house, at any time of day. Lesser Whistling Teals used to fly every evening at seven over my house in Begum Bazaar, perhaps from Miralam Tank to Hussain Sagar. And if their characteristic whistles, uttered on the wing, reached my ears, my heart leapt out of the confines of the room I sat in and thrilled in the flock of birds winging their way to roost through the gathering dark, whistling as they went.

Begin birding by ear today. Birding will never be the same again.

44

My Kind of Birding

In recent years, I have often heard a refrain that one sees the same species of birds on field trips. Initially, I would get cross at such an observation, which was remiss of me, for everyone does not think the way I do, and that's okay. Later, I was bemused, for how could new species be ordered up for a birder? Now I am concerned, for I think this is a symptom of a deeper malady.

I have been birding for a little over three decades, and never have I been put off by the prospect of seeing the same common species on a birding outing. When something new turned up during an outing or a rare bird was sighted, we were ecstatic. The excitement ignited a sense of awe in the entire group. But I cannot recall even one instance of a birder complaining about a lack of species novelty. So, this lament is new to my ears.

I feel that it is a symptom of a syndrome often called boredom, which is particularly prevalent in the digital generation, for whom the predilection to channel-surf through life's situations is an overpowering need, and this gives them a false sense of control over their lives, of the desire to change a situation at the mere press of a button. The downside of this is an ever-decreasing attention span. If this phenomenon begins to seep into one's very nature, then activities that are not considered 'essential' to the life of a person, say, leisure, inevitably fall prey to a ceaseless, futile hunger for novelty in whatever one does. Given such a bent of mind, and with the mind-bending peer pressures rampant today, how long will it take, I often ask myself, before this malady leaches into the more essential fabric of one's life? I see signs of it daily in traffic snarls. Tempers flare at the avoidable inconvenience, but people do not hesitate to slip out of their lane on spotting a chance for a quick exit. If everyone stuck to traffic rules, the flow would undoubtedly be smoother.

Leisure activities, as I was saying, are the first victim of such aberrant behaviour. I am of the firm belief that the constant desire to spot at least one species never seen before, and the paradoxical moaning that follows the no-show of a new species, is a symptom of the channel-surfing mindset. I am concerned about such an overpowering desire for novelty, when a dynamic activity like birding becomes suspect of being mired in stagnation, and the disgruntled birder begins to wonder whether it would be worthwhile to spend another mundane morning tramping through the countryside, chasing the same old birds!

Birdwatching is not an eternal quest for rarity, though no birder denies the thrill of sighting one. It is not about racing all over the landscape and tallying a century of birds before lunch, though no birder will deny the special joy of such a 'ton'. It is not twitching for the greatest number of species seen, though there have been many that have basked in the sunshine of that self-indulgent high life (they truly miss the wood for the trees).

Birdwatching, in its essence, is the fine art of becoming invisible – of merging into the surroundings in such a way that the breath which nature has held back upon your entry into its parlour is joyfully exhaled and normal respiration restored; in such a way that the frozen statues of animate wildlife, interrupted by your brashness, are coaxed into resuming their activities; in such a way that your aural and visual senses are drenched with the buoyancy of life; in such a way that you find a way outside yourself and become a part of the pageant around you.

This does require the cultivation of a patience that slows down your pace to that of the elemental cycles dominating the flow of life in an immaculate world run entirely without human help. It requires the marshalling and realigning of vision and a new focus of hearing so that you absorb every single sound and identify its source and gradually its nuances, its cadenzas. It demands a preoccupation with stillness.

What are the rewards of this exercise? I can think of at least two that will last you a lifetime. One, you will begin to notice things about your surroundings that you never knew existed, bringing you immediate, immeasurable joy. Two, your restless inner dynamo will wind down to such an extent that you discover a quietude, a stillness within you; a fount for a fresh view of your surroundings, a new approach to life based on re-energized sensitivities.

In the field, this approach opens the doors of a new world to your mind's eye. You remain standing in front of a fruiting neem, while the larger group of birders moves on, having identified one or two species. You inhale the aroma of its foliage. A fluty whistle from its canopy leads your eye to the Iora. He is dressed in breeding regalia – a jet cap and coat, deep canary shirt, white epaulettes. He courts a hen Iora with song, he postures, he patrols; his aria persuades her and dissuades rivals from his territory. You do not exist in his scheme of things, just in your own sensory world, as a witness. You stay with him as he perambulates through the leafy canopy, lifting his warbling beak skyward, fluffing his velvet beret, vibrating his dark tail, standing on tiptoe in the fervour of his operatic song. You are trapped by its intensity. You are mesmerized by its elemental simplicity, by the realization that the Iora's entire world, in that moment, is its song, and that he has enmeshed you in it, albeit momentarily, till your focus expands to take in the larger picture – the shining curved leaves of the neem, the soft yellow fruit pods, the darkened bark, damp from last night's rain, the tangled undergrowth.

You spy a movement from the corner of your eye and realize it is an Ashy Prinia that's flown into the Iora's neem, and you stay with it. As your senses expand, your absorption of the drama around you becomes acute and before you are aware of it, you are invisible to yourself, a part of the very landscape you've come to partake of, all eyes and ears, inhaling its scents, feeling it on your skin. If you become aware of yourself, the spell is broken and the pageant melts away into simple, mundane, two-dimensionality. The trick is to make yourself invisible, while remaining completely present in your surroundings.

Now you are on the path to my kind of birding.

45

From Analog to Digital

When I began birding, in 1978 or thereabouts, only two tools were considered essential for a birdwatcher: a pad and a pencil. Even binoculars and bird guide were sidelined as superfluous. The emphasis was on observing and noting down what you saw in the field itself. A lapse of time between observing and writing down, relying on memory, conjured 'thingamajigs'. You relied on your eyes and ears to imbibe and your digits to record. In the process, there were no distractions, no intrusions. You connected with your subject in a way that left indelible impressions, not just as an experience in the mind, but also as a factual record on paper. Later, at home, or in a library, when you read up your notes and compared them with what was known about the species to the scientific world, you realized perhaps that what you had recorded that morning was common knowledge, and you took heart in knowing you had reconfirmed a fact, that your notes said the same things as Sálim Ali's *Book of Indian Birds* or Whistler's *Popular Handbook of Indian Birds*.

More often, you realized you'd missed recording stuff they'd noticed, and yearned to go back there soonest and see or hear for yourself. But the best was when you'd noticed what no one else had. It was electrifying and beset you with self-doubt, but then your notes supported you. So, you wrote to an expert, or a more experienced friend, for advice and waited for a response. Invariably, they urged you to write up your notes and send them for publication to one of the two extant journals, the *Journal of the Bombay Natural History Society* or the *Newsletter for Birdwatchers*. Once published, you too sported a feather in your cap!

If you had the inclination, and the time, you could fair out your notes at home. My system was a tad elaborate. I typed out the notes from each trip on foolscap sheets, bannered with the name of the place, date, and additional notes like weather, habitat, names of other birdwatchers (OBWs), etc. These were filed away alphabetically. I numbered all my field notebooks and their pages. These numbers were also typed on the relevant trip list sheet, all of which helped in quick retrieval of data later. Typing on single sheets allowed their alphabetical assimilation, which was useful when one wished to see the chronological notes of a particular place, all being in one place, not scattered over several notebooks.

I went a step further and created sheets for each species

I saw and noted in long hand the date, place of observation, notes, and reference number of the field notebook. These I filed by the corresponding volume of Sálim Ali and Dillon Ripley's magnum opus, *Handbook of the Birds of India and Pakistan* and not alphabetically. It was tedious, but immensely rewarding, e.g., all my Grey Wagtail records were on a single sheet of paper or more, all in one place.

When the digital era created ways of storing these analog records in the new format and promised innumerable ways of data retrieval and transfer, in 1995, I migrated all my data to a commercial database software called 'Bird Recorder'. I had read about it in a copy of the *OBC Bulletin*, and the software was developed by Jack Levene's Wildlife Computing Limited. But my notebooks have remained my prized possessions, with their quirky sketches, irreverent observations, on-the-spot jottings of banter, shayari, etc.

As my notes grew, even if they comprised mundane lists of birds, they were data that could be used one day. But it frustrated me no end that I was unable to make them available to a wider audience. I did not want it sitting locked inside my desk. If that was to be their fate, their usefulness would be only partially served – by being personally gratifying to me.

But the launch of eBird changed all that. Here, finally, was an opportunity to add my life list to a database that truly represented my aspirations for my birding notes. The transfer from Bird Recorder to eBird was a breeze through the interface. Now all my notes are on eBird in the public domain. To my mind, they could not have been put to better use than form a small part of the vast numbers of records comprising eBird.

I had also entered historical published records, such as Sálim Ali's *Hyderabad Survey*, into Bird Recorder, and hoped to key in the contemporary records too, but that remains undone. The idea was to create a comprehensive digital database for the erstwhile political state of Andhra Pradesh. Now that scanned versions of journals and PDFs are so easily available, I hope someone will take up the challenge and complete the work.

Despite the convenience of eBird, even that of using it directly in the field, nothing can replace the original tools of a birdwatcher, for the activity is not merely an insipid listing of birds. Everyone should discover for themselves the character of every species, the habitat it lives in, the miraculous change of seasons, the trigger that clicks in place instinct. After a while, eBird notes become standardized in form and output, whereas when the emotions you experience while watching an avian drama in the wild are translated in your own hand into your notebooks, they thrill you throughout your lifetime.

46

Intimacy Without Violation:
A Short Meditation on Bird Photography[40]

It is a common scene in birding groups today: people with long-lens cameras outnumber those with binoculars. Both types of optics are mandatory equipment for the hobby; we *need* the power of these magnifying tools to bring the birds closer to us because whether we pursue them through urban, rural or wild landscapes, we intrude into realms that have become increasingly alien to us (as we have to them) because of our ever-deepening withdrawal into our buzzing cities.

In that, the art of birding becomes an act of voyeurism. Birders unabashedly ogle at the way birds live their lives, irrespective of what they're doing, and insensitive to the effect this behaviour might have upon them. Just step back and watch the birds, and you'll realize they are watching us.

All the while they watch us, their attention is removed from their surroundings. In nature, this momentary distraction could be tragic at its worst or energy depleting at its most benign. When their guard is down, death swoops in disguised as a predator. When their energy is expended in abnormal activity, replenishing it requires extra work which is fraught with its attendant dangers.

The cardinal rule of birding is that the bird comes first – not the good view that a birder desires, not the correct identification, not the best photographic shot. With the steady improvement of optical tools, we are fortunately better equipped to view/photograph birds from a distance. Yet, there are people who are not easily satisfied and want better views or angles or a decrease in proximity with their subject. This is a trait more often seen among bird photographers (rather than 'simple' birders). While bin-toting birders are happy to watch from a distance, aware that a closer approach might scare their quarry, those behind rapidly firing telecameras are not so punctilious. Harnessing their stalking skills, they sidle closer and closer for the one shot that will earn them plaudits in the digital world: the catch-light winking in the eye, the gore-soaked talon, the eviscerated intestines, the ritual of feeding, the frenetic display of courtship – the 'Aha!' moment. In their pursuit, they kill two birds with one stone – to use an unfortunate pun – the pleasure of bin-toting birders, and the birds themselves, compromising, in the latter act, the integrity of the cardinal rule.

So, you, the prospective bird photographer, must decide between being a tourist, snapping birds (yes, even through those huge lenses), or a bird photographer worthy of that moniker. If it is the former – which has its own brilliant uses – then follow a tourist's and birder's codes.[41] Keep your distance from your subject, keep the sanctity of the place (or the sense of safety in the bird) intact, and allow fellow tourists or birders an equal opportunity to admire and absorb the moment. Bring snapshots to help identify the bird or to fix memory. If you get one or more photograph from a distance, be satisfied. If not, don't pursue the bird for a better shot, a better angle, better light, etc. Don't become a tourist trespassing into restricted precincts, or a birder walking on freshly planted fields, or through grasslands harbouring ground-nesting species. Don't vandalize with graffiti or impoverish by collecting souvenirs or endanger the equilibrium of a bird's life with your insensitivity. Care enough for the bird and ensure you do not scare it into flight. If it is unidentifiable in your photo, let the mystery remain. Isn't that part of birding too?

I believe that if you end up looking at birds only through a viewfinder, you miss the bigger picture of the drama, the wood for the trees as it were, and you are left with an education that is only rudimentary. The view is not so myopic when you birdwatch through binoculars. Try it.

But if you wish to become a bird photographer, then this advice comes from one of the best photographers in the country, Dhritiman Mukherjee. Leave behind your cameras on your first photographic foray. Take only your binoculars. Pursue your subject for lengths of time, study it deeply: its movements relative to its surroundings and to the movement of light and shade; its behaviour in different situations; its modes of interactions with other wildlings; its favourite perches. Stay with it from dawn to dusk, unobtrusively watching it all the time. Become intimate with its life cycles, without violating its sense of space or its sense of safety. Once you've done this, leave your binoculars and pick up your camera. By becoming a familiar of the birds, you would have learnt how to tackle their shy, timid flightiness. You'd have deciphered the contours of the land relative to its presence and chosen the spots for the photographs you want to make. If need be – and it is often so – you would have devised the subterfuge required to disguise your human form from the bird, becoming, as Aldo Leopold yearned, 'a muskrat, eye-deep in the marsh',[42] when the geese return; or the scarecrow amidst the shivering crop upon which a petrified lark perches, frozen below a hovering harrier; or a hay-laden bullock cart near a displaying florican; or a tree trunk at the edge of a forest clearing. You would have divined the bird's movements and those of the light. You would have allowed the land to become intimate with you. Above all, you would have formed a peripheral instinct for the pulse of life in that pocket of wilderness. None of this would have been possible had you continued to violate the brittle balance of life that permeates wildernesses. Now the bird is no longer startled by your presence, and when it sees you, instead of taking frantic, evasive action, it holds you in its gaze, not as an aberration, but as a part of its birdscape.

When you begin to shoot, you'll be thrilled with your pictures not because they are good, but because the bird is not watching you, it is living its life.

47

When You Finally Decide to Publish Your Birding Notes!

There will invariably come a juncture in the life of a birdwatcher when the desire to record, for posterity, what has been seen and experienced in the field becomes almost overpowering, and that enthusiastic moment, when the observation gets transformed into the black-and-white cement of print, is vital for ornithology, for the note will now join the library that forms the rank and file of recorded bird study.

But when this journey of the observation, from the mind's eye to the pen hovers over a blank sheet of paper, it does pay to pause and ponder the very process. What should comprise the framework of the note?

You would have some written notes about the encounter in your field notebook – these should form the very backbone of your paper. If you did not commit to paper or to pixel what you thought was unique and worthy of reporting, then you are on slightly shaky ground, for memory is truly a flighty bird in the bush. So always jot down as many details as you possibly can on the spot. Try and sketch the bird if a photographic record is not possible. Do not bother about it being pretty, but always aim for accuracy of description. These are your primary data and the very bedrock of your tryst with the printed page.

Once you are convinced that you have something novel to share with fellow ornithologists, you need to bolster that conviction with historical support, for science is indeed the process of building upon earlier blocks, with the inviolable premise of general integrity being the primary ethic of the ornithologist. Look through published literature for more information on your theory or observation. Assess the past objectively, logically, dispassionately and weigh in the balance of fact versus 'flight of fancy' whether what you thought was new is truly so. Ponder if what you have observed substantially adds to the sum of knowledge. Does it provide any fresh insights? These are the points you will have to ruminate over. Perish the thought of publishing if it is not so, treating the exercise not as one in futility, but as an educative one.

But if you *do* have something, then zoom out and

observe the entire overall picture to reassess how your unique observation fits in the larger jigsaw puzzle. Once you are convinced, plunge in and try to go to the source of every available and relevant past record. If this entails delays and tribulations, so be it, for fruit always tastes best when ripe. Summarize clearly and cogently what you have researched, for you must cite references in the appropriate places to support your marshalled data. (The citations should correspond to an alphabetically, chronologically arranged bibliography listed under a references section at the end of your paper.) Do not fear if you plough up more of the field than required, for at least you will have the satisfaction of not leaving any stone unturned. Now, separate the cheese from the chalk and prudently select only the most relevant references for your paper.

Conclude with a succinct discussion about your theory or observation to arrive at what you have deduced. This is your framework, a foundation based on fact, supported by historical documentation and a discussion that positively clinches your point of view.

Your work is done, or so you presume. But it is only half the battle won. You need to study the 'house style' of your target journal. By this I mean begging the indulgence and presumption of the wizened, the way the journal crosses its t's and dots its 'i's; the way it uses capitalization; the lay of geographical terms; the style of citations; the precise method of the reference section; the taxonomic style; the English and scientific nomenclature it subscribes to, and so forth.

The best way of doing this is to look up any guidelines to

authors that the journal might carry within its pages or on its website. You must also try and get a previously printed issue of the journal to help you to study the style of articles published within its covers.

When you have legitimate facts, have completed your homework, and have fully understood the house style, you are finally ready to put pen to paper. When it's done, sleep over what you've written. Review it afresh. Show it to your peers and professors. Encourage constructive criticism with a view to improvement. Badger them to assess the scientific rigour and to comment on language, for your aim is to write in a way that conveys a primarily scientific or semi-scientific subject to a lay, non-scientific readership with both simplicity and accuracy. Trust me when I say that your harshest critic at this stage is your best friend compared to the talons that will inevitably slash your note in the peer review, or worse, the beaks of sharp readers that will slice you like a thrush dispatching a worm.

A paper that is authentic, that has veracity, that has followed the house style, and whose language is clear and simple, immediately attracts the attention of the perpetually harassed editor, who picks it from the morass of misguided manuscripts littering their desk, joyously recognizing your efforts in making your work what it is and sends it off to one or more peer reviewers – that tribe of gurus who can be either the bane or the benediction of a manuscript – relegating it to the dreaded 'TO DO' pile if it is tardy, or fervently pouncing on it with zealous red-inked pens if it is not.

Revere the bloodstained shreds of your manuscript when

they finally return from their dreaded baptism by editorial fire. Dress all its wounds carefully, meticulously, and lovingly – snipping, stitching, applying salve. Leave no wound unattended, for it will surely fester till you are forced to nurse it back to health.

After a mandatory period of convalescence, when you sleep over your handiwork, revisit it once again, calmly and dispassionately, before marching it off once again to the editor. When, finally, the printed journal is in your hands, and you see only one name in its many pages, the trauma of labour is instantly forgotten and you, once again, take up your mighty pen.

Glossary

Antediluvian Belonging to a time before the great Biblical Flood

Banding Putting a metal ring with a unique number on a bird's leg for identification purposes

Carpal joint On a bird's wing, the joint between the arm and the hand

Coverts Feathers covering the base of flight and tail feathers, both on the upperside and underside

Disc A circlet of feathers around the face

Falcon A falconry term for a female falcon; also used as a generic term for falcons

Gape The mouth

Gizzard Stomach

Gular pouch A pouch between the lower mandible and the upper throat

Gullet The anterior part of the food channel, or upper oesophagus

Mandibles Two parts of the bill. Often used strictly for the lower part

Moult The growth of new feathers, usually through a process of regular renewal

Nidicolous Referring to a nestling hatched naked and helpless

Oology The study or collecting of birds' eggs

Pelage Fur, hair, or wool; used here for covering or skin

Pinions Wing/s

Precocious Chicks hatching in a fairly advanced stage, with open eyes, and able to walk or run within minutes or hours

Primary *See* primaries

Primaries Outer flight feathers attached to the hand of a bird. The singular is primary.

Ringing *See* banding

Supercilium Eyebrow

Secondary *See* secondaries

Secondaries Inner flight feathers attached to the forearm of a bird's wing

Syrinx The vocal organ of a bird

Tarsus Lower leg, or foot

Tiercel A falconry term for a male falcon

Umwelt The world as it is experienced by a particular organism: the worlds they perceive, their Umwelten, are all different

Ventral Referring to the lower part of the body

Xerophytic Plants needing very little water

Non-human Forms of Life Mentioned in the Text

(With Alternative Names Within Parenthesis)

Arthropods

Blue Bottle *Calliphora vomitoria* (Blue Bottle Fly)

Danaid Eggfly *Hypolimnas misippus*

Jezebel *Delias eucharis* (Common Jezebel)

Tiger Beetle Carabidae: Cicindelinae

Birds

Ashy-crowned Finch Lark *Eremopterix grisea* (Ashy-crowned Sparrow Lark)

Ashy Prinia, *See* Ashy Wren-Warbler

Ashy Wren-Warbler *Prinia socialis* (Ashy Prinia)

Avocet *Recurvirostra avosetta* (Pied Avocet)

Barn Owl *Tyto alba* (Common Barn Owl)

Barn Swallow *Hirundo rustica* (Common Swallow)

Barred Jungle Owlet *Glaucidium radiatum* (Jungle Owlet)

Baya *Ploceus philippinus* (Baya Weaver)

Black Bulbul *Hypsipetes ganeesa* (Square-tailed Bulbul)

Black Drongo *Dicrurus macrocercus* (King Crow)

Black Eagle *Ictinaetus malaiensis*

Black Kite *Milvus migrans* (Hindi/Urdu: cheel)

Black-capped Babbler *Dumetia atriceps* (Black-fronted Babbler)

Black-crested Bulbul *Rubigula gularis* (Flame-throated Bulbul)

Black-headed Cuckoo-Shrike *Lalage melanoptera* (Black-headed Cuckooshrike)

Black-headed Myna *Sturnia pagodarum* (Brahminy Starling)

Black-headed Oriole *Oriolus xanthornus* (Black-hooded Oriole)

Black-shouldered Kite *Elanus caeruleus* (Black-winged Kite)

Black-tailed Godwit *Limosa*

Black-winged Stilt *Himantopus*

Blossom-headed Parakeet *Psittacula cyanocephala* (Plum-headed Parakeet)

Blue Rock Pigeon *Columba livia* (Feral Rock Pigeon)

Blue Rock Thrush *Monticola solitarius*

Bluethroat *Luscinia svecica*

Blue-headed Rock Thrush *Monticola cinclorhyncha* (Blue-capped Rock Thrush)

Bonelli's Eagle *Aquila fasciata*

Brahminy Duck *Tadorna ferruginea* (Ruddy Shelduck)

Brahminy Kite *Haliastur indus*

Brahminy Myna *Sturnia pagodarum* (Brahminy Starling)

Brainfever Bird *Hierococcyx varius* (Common Hawk Cuckoo)

Bronzed Drongo *Dicrurus aeneus*

Brown-breasted Flycatcher *Muscicapa muttui*

Brown Shrike *Lanius cristatus*

Butastur teesa White-eyed Buzzard

Cattle Egret *Bubulcus ibis*

Chestnut-headed Bee-eater *Merops leschenaulti*

Chestnut-shouldered Petronia *Gymnoris xanthocollis* (Yellow-throated Sparrow)

Cliff Swallow *Petrochelidon fluvicola* (Streak-throated Swallow)

Common Babbler *Argya caudata*

Common Iora *Aegithina tiphia*

Common Myna *Acridotheres tristis*

Common Snipe *Gallinago*

Common Swallow *See* Barn Swallow

Common Woodshrike *Tephrodornis pondicerianus*

Coppersmith *Psilopogon haemacephalus* (Coppersmith Barbet)

Coucal *Centopus sinensis* (Greater Coucal)

Crested Hawk-Eagle *Nisaetus cirrhatus* (Changeable Hawk Eagle)

Crimson-throated Barbet *Psilopogon malabaricus* (Malabar Barbet)

Cuckoo, Common Hawk *See* Brainfever Bird

Curlew *Numenius arquata* (Eurasian Curlew)

Curlew Sandpiper *Calidris ferruginea*

Dabchick *Tachybaptus ruficollis* (Little Grebe)

Cursorius bitorquatus See Jerdon's Courser

Darter *Anhinga melanogaster* (Oriental Darter)

Dunlin *Calidris alpina*

Dusky Crag Martin *Ptyonoprogne concolor*

Eastern Skylark *Mirafra gulgula* (Oriental Skylark)

Egyptian Vulture *Neophron percnopterus*

Emerald Dove *Chalophaps indica* (Asian Emerald Dove)

Eurasian Wigeon *Mareca penelope*

Gold-fronted Chloropsis *aurifrons* (Gold-fronted Leafbird)

Golden Oriole *See* Indian Golden Oriole

Golden-backed Woodpecker *Dinopium benghalense* (Black-rumped Flameback)

Goshawk *Accipiter gentilis* (Northern Goshawk)

Glossy Ibis *Plegadis falcinellus*

Great Crested Grebe *Podiceps cristatus*

Great Indian Bustard *Ardeotis nigriceps* (Indian Bustard)

Greater Flamingo *Phoenicopterus ruseus*

Greater Racket-tailed Drongo *Dicrurus paradiseus*

Greenish Leaf Warbler *Phylloscopus trochiloides* (Greenish Warbler)

Greenshank *Tringa nebularia* (Common Greenshank)

Grey Drongo *Dicrurus leucophaeus* (Ashy Drongo)

Grey Heron *Ardea cinerea*

Grey Hornbill *Ocyceros birostris* (Indian Grey Hornbill)

Grey Junglefowl *Gallus sonneratii*

Grey Partridge *Ortygornis pondicerianus* (Grey Francolin)

Grey-headed Myna *Sturnia malabarica* (Chestnut-tailed Starling)

Grey-necked Bunting *Emberiza buchanani*

Hair-crested Drongo *Dicrurus hottentottus*

Himalayan Whistling Thrush *Myophonus caeruleus* (Blue Whistling Thrush)

House Crow *Corvus splendens*

House Swift *Apus affinis* (Little Swift)

Indian Courser *Cursorius coromandelicus*

Indian Cuckoo *Cuculus Micropterus*

Indian Eagle Owl *Bubo bengalensis*

Indian Golden Oriole *Oriolus kundoo*

Indian Pitta *brachyura*

Indian Robin *Copsychus fulicatus*

Indian Skylark *Alauda gulgula* (Oriental Skylark)

Jacanas Jacanidae

Jerdon's Bushlark *Mirafra affinis*

Jerdon's Courser *Rhinoptilus bitorquatus* (old scientific name: *Cursorius bitorquatus*)

Jungle Babbler *Argya striata*

Jungle Crow *Corvus macrorhynchos* (Large-billed Crow)

Jungle Myna *Acridotheres fuscus*

Kestrel *Falco tinnunculus* (Common Kestrel)

Koel *Eudynamys scolopaceus* (Asian Koel)

Lammergeier *Gypaetus barbatus* (Bearded Vulture)

Large Cuckooshrike *Coracina macei* (Large Cuckooshrike)

Large Green Barbet *Psilopogon zeylanicus* (Brown-headed Barbet)

Large Grey Babbler *Argya malcolmi*

Large Pied Wagtail *Motacilla maderaspatensis* (White-browed Wagtail)

Lesser Whistling Teal *Dendrocygna javanica* (Lesser Whistling Duck)

Little Egret *Egretta garzetta*

Little Ringed Plover *Charadrius dubius*

Little Stint *Calidris minuta*

Magpie-Robin *Copsychus saularis* (Oriental Magpie Robin)

Malabar Trogon *Harpactes fasciatus*

Marsh Harrier *Circus aeruginosus* (Western Marsh Harrier)

Median Egret *Ardea intermedia*

Monarch Flycatcher *Hypothymis azurea* (Black-naped Monarch)

Montagu's Harrier *Circus pygargus*

Mottled Wood Owl *Strix ocellata*

Nightingale *Luscinia megarhynchos*

Nightjar *Caprimulgus* species

Olive-backed Pipit *Anthus hodgsoni*

Orange Minivet *Pericrocotus flammeus*

Osprey *Pandion haliaetus*

Ostrich *Struthio camelus* (Common Ostrich)

Paddy Bird *Ardeola grayii* (Indian Pond Heron)

Painted Partridge *Francolinus pictus* (Painted Francolin)

Palm Swift *Cypsiurus balasiensis* (Asian Palm Swift)

Paradise Flycatcher *Terpsiphone paradisi* (Indian Paradise-flycatcher)

Parakeet *Psittacula* species

Peafowl *Pavo cristatus* (Indian Peafowl)

Peninsular Scimitar Babbler *Pomatorhinus horsfieldii* (Indian Scimitar Babbler)

Peregrine Falcon *Falco peregrinus*

Pied Bushchat *Saxicola caprata*

Pied Myna *Gracupica contra* (Asian Pied Starling)

Pied-crested Cuckoo *Clamator jacobinus* (Pied Cuckoo)

Pygmy Woodpecker *Yungipicus nanus* (Brown-capped Pygmy Woodpecker)

Pipits *Anthus* species

Plaintive Cuckoo *Cacomantis passerinus* (Grey-bellied Cuckoo)

Pompadour Pigeon *Treron affinis* (Grey-fronted Green Pigeon)

Pond Heron *See* Paddy Bird

Purple Heron *Ardea purpurea*

Purple Moorhen *Porphyrio poliocephalus* (Grey-headed Swamphen)

Purple Sunbird *Cinnyris asiaticus*

Purple-rumped Sunbird *Leptocoma zeylonica*

Quaker Babbler *Alcippe poioicephala* (Brown-cheeked Fulvetta)

Red Spurfowl *Galloperdix spadicea*

Red-crested Pochard *Netta rufina*

Red-headed Falcon *See* Red-headed Merlin

Red-headed Merlin *Falco chicquera* (Red-necked Falcon)

Red-rumped Swallow *Cecropis daurica*

Red-vented Bulbul *Pycnonotus cafer*

Red-wattled Lapwing *Vanellus indicus*

Red-whiskered Bulbul *Pycnonotus jocosus*

Red-winged Bushlark *Mirafra erythroptera* (Indian Bushlark)

River Tern *Sterna aurantia*

Roller *Coracias benghalensis* (Indian Roller)

Rose-ringed Parakeet *Psittacula krameria*

Rosy Pastor *roseus* (Rosy Starling)

Ruff *Calidris pugnax*

Rufous-bellied Babbler *Dumetia hyperythra* (Tawny-bellied Babbler)

Rufous-tailed Finch-lark *Ammomanes phoenicura* (Rufous-tailed Lark)

Sarus *Antigone antigone* (Sarus Crane)

Scimitar Babbler *Pomatorhinus horsfieldii* (Indian Scimitar Babbler)

Shaheen Falcon *Falco peregrinus* subspecies

Shama *Copsychus malabaricus* (White-rumped Shama)

Shikra *Accipiter badius*

Short-eared Owl *Asio flammeus*

Short-toed Eagle *Circatus gallicus* (Short-toed Snake Eagle)

Shoveler *Spatula clypeata* (Northern Shoveler)

Small Green Bee-eater *Merops orientalis* (Green Bee-eater)

Small Indian Pratincole *Glareola lactea* (Small Pratincole)

Sparrow *Passer domesticus* (House Sparrow)

Sparrowhawk *Accipiter nisus* (Eurasian Sparrowhawk)

Spotted Babbler *Pellorneum ruficeps* (Puff-throated Babbler)

Spotted Owlet *Athene brama*

Spotted Sandpiper *Tringa glareola* (Wood Sandpiper)

Sykes's Crested Lark *Galerida deva* (Sykes's Lark)

Tailor Bird *Orthotomus sutorius* (Common Tailorbird)

Temminck's Stint *Calidris temminckii*

Tickell's Blue Flycatcher *Cyornis tickelliae*

Tickell's Flowerpecker *Dicaeum erythrorhynchos* (Pale-billed Flowerpecker)

Treepie *Dendrocitta vagabunda* (Rufous Treepie)

Turkey Vulture *Cathartes aura*

Verditer Flycatcher *Eumyias thalassinus*

Wagtails *Motacilla* species

White-eyed Buzzard *Butastur teesa*

White Stork *Ciconia ciconia*

White-backed Vulture *Gyps bengalensis* (White-rumped Vulture)

White-bellied Drongo *Dicrurus caerulescens*

White-bellied Treepie *Dendrocitta leucogastra*

White-breasted Kingfisher *Halcyon smyrnensis* (White-throated Kingfisher)

White-browed Bulbul *Pycnonotus luteolus*

White-eye *Zosterops palpebrosus* (Indian White-eye)

White-headed Babbler *Argya affinis* (Yellow-billed Babbler)

White-spotted Fantail Flycatcher *Rhipidura albogularis* (Spot-breasted Fantail)

White-throated Ground Thrush *Geokichla citrina* (Orange-headed Thrush)

Yellow-cheeked Tit *Machlolophus aplonotus* (Indian Black-lored Tit)
Yellow-browed Bulbul *Acritillas indica*
Yellow-eyed Babbler *Chrysomma sinense*
Yellow-throated Bulbul *Pycnonotus xantholaemus*

Fish

Mahseer *Tor remadevii* (Hump-backed Mahseer)

Mammals

Blackbuck *Antilope cervicapra*
Black-naped Hare *Lepus nigricollis* (Indian Hare)
Bonnet Macaque *Macaca radiata*
Hyena *Hyaena* (Striped Hyena)
Indian Flying Fox *Pteropus medius*
Indian Fox *Vulpes bengalensis* (Bengal Fox)
Indian Giant Squirrel *Ratufa indica*
Indian Wolf *Canis lupus*
Mongoose *Urva edwardsii* (Indian Grey Mongoose)
Pipistrelle *Pipistrellus* species
Sambhar *Rusa unicolor*
Squirrel *Funambulus* species
Tiger *Panthera tigris*

Plants

Babool *Acacia* species
Banyan tree *Ficus benghalensis*

Beddome's Cycad *Cycas beddomei*
Ber *Ziziphus* species
Camel's Foot Climber *Vanera vahlii* (Bauhinia Climber)
Cattails *Typha* species
Cockscomb *Celosia* species
Coconut Palm *Cocos nucifera*
Coral Tree *Erythrina variegata* (Indian Coral Tree)
Fishtail Palm *Caryota urens*
Flame of the Forest Tree *Buteo monosperma*
Glory Lily *Gloriosa superba*
Henna *Lawsonia inermis* (Hindi: *Mehendi*)
Millingtonia hortensis (Indian Cork Tree)
Morinda Tree *Morinda tinctoria*
Palmyra Palm *See* Toddy Palm
Peepal Tree *Ficus religiosa*
Prosopis species
Mango Tree *Mangifera indica*
Sheesham Tree *Dalbergia sissoo*
Silk Cotton Tree *Bombax ceiba*
Sitaphal *Annona squamosa* (Custard Apple)
Tamarind Tree *Tamarindus indus*
Toddy Palm *Borassus flabellifer* (Palmyra Palm)
Tridax *Tridax procumbens* (Coatbuttons/Tridax Daisy)
Wild Sesame *Sesamum indicum*

Reptiles

Flying Lizard *Draco dusummieri* (Indian Flying Lizard)
Garden Lizard *Calotes versicolor* (Oriental/Common Garden Lizard)

Record of Previous Publication

Some of the essays in this book have been previously published in newspapers, ornithological journals, and blogposts. The versions that appear in this book may have been updated and edited. A list is enclosed below. Text within square brackets is the original published title.

1. Pittie, Aasheesh. 'Tryst with Jerdon's Courser *Cursorius bitorquatus* (Blyth)'. *Newsletter for Birdwatchers* 39, no. 6 (1999): 83–84.

2. ———. 'Can We Afford to Lose the Great Indian Bustard?' *Hornbill* (April–June 2001): 24–26.

3. ———. 'Silence of the Sparrows'. *The Hindu (MetroPlus)*. 10 December 2002.

4. ———. 'The Death of a Nightjar'. *Newsletter for Birdwatchers* 44, no. 1 (2004): 8–9.

5. ———. 'Falcons in Focus'. *Newsletter for Ornithologists* 1, nos. 1 and 2 (2004): 30–31.

6. ———. 'Friends from Wingdom'. *Weekend Newstime*. 5 August 1984.

7. ———. 'Disappearing Dabchicks'. *Weekend Newstime*. 10 June 1984.

8. ———. 'The Song of the Iora'. *Sunday Chronicle*. 11 December 1983.

9. ———. 'The Kotwal of the Countryside [Tale of a Long-tailed Small Bird]'. *Sunday Chronicle*. 6 February 1984.

10. ———. 'Master of the Urban Skies'. *Sunday Chronicle*. 25 December 1983.

11. ———. 'Vulture Culture: Not All That "Offal"'. *Express Weekend*. 25 June 1988.

12. ———. 'Owl Therapy'. *Andhra Pradesh Times*. 28 November 1997.

13. ———. 'Spotted Owlet: The City Gossip [Does This Bird Gossip?]'. *Deccan Chronicle (School Chronicle)*. 1 November 1999.

14. 'A Coppersmith in a Tree [Tuk tuk! Look I'm here]'. *Deccan Chronicle (School Chronicle)*. 4 October 1999.

15. ———. 'Did-you-do-it?' *Deccan Chronicle*. 1999.

16. ———. 'One, Two, Three … Ten, Twenty, Thirty'. *Hornbill* no. 2 (1987): 7–9.

17. ———. 'End of the day birding …' *Pitta* 50 (1995): 1.

18. ———. 'Bird on a Wire [Since the World Got Wire, Birds Have Found a New Perch …]'. *Pitta* 52–54 (1995): 1.

19. ———. 'Revisiting Shamirpet Lake [Shamirpet Lake]'. *Pitta* 69 (1997): 1–2.

20. ———. 'Remembrances of Birdsongs Past'. *Andhra Pradesh Times*. 9 November 1997.

21. ———. 'Uma Maheshwaram: Sacred Indeed [Field trip to Uma Maheshwaram, 22 November 1998]'. *Pitta* 93 (1999): 1–2.

22. ———. 'A Bird in Hand: The Patience and Persistence of a Bird Ringer. [Bird Banding in the Sri Venkateshwara National Park]'. *Sanctuary Asia* 19, no. 6 (1999): 56–59.

23. ———. 'Reflections on a Short Trek to Tala Káveri [Trekking to Tala Káveri]'. *Sanctuary Asia* 20, no. 6 (2000): 34–38.

24. ———. 'A Personal IBA [Personal and National IBAs]'. *Newsletter for Ornithologists* 1, no. 6 (2004): 81.

25. ———. 'Mini Doses of Life for Harried Souls'. *Deccan Chronicle*. 26 March 2000.

26. ———. 'Beacons of Brilliance'. *Deccan Chronicle (School Chronicle)*. 21 February 2000.

27. ———. 'Seeing the Spectacular in the Ordinary [Man Can Create Glorious Gardens]'. *Deccan Chronicle (School Chronicle)*. 17 September 2000.

28. ———. 'Humayun Abdulali, 1914–2001'. *OBC Bulletin* 34, (December 2001): 8.

29. ———. 'S is for Shikra'. *Pitta* [New Series] 12, no. 6 (2015): 1–6.

30. ———. 'Remembering Zafar Futehally (1920–2013)'. Nature in Focus (blog). 18 August 2017. https://www.natureinfocus. in/environment/remembering-zafar-futehally-1920-2013.

31. ———. 'Checklists: Ticking it off [Checklists – The Birds of Andhra Pradesh]'. *Hornbill* no. 2 (1990): 4–7.

32. ———. 'Birding by Ear [Birding by Ear Is as Enjoyable as Bird-watching! …]'. *Pitta* 48 (1994): 1.

33. ———. 'My Kind of Birding'. *Bird Watcher's Digest*. November/December 2013. 42–45.

34. ———. 'From Analog to Digital'. Bird Count India (blog). 20 April 2017. https://birdcount.in/analog-to-digital/.

35. ———. 'Intimacy Without Violation'. Nature in Focus (blog). 14 May 2018. https://www.natureinfocus.in/environment/intimacy-without-violation.

Notes

Birds

1. Zafar Futehally, *Newsletter for Birdwatchers* 39, no. 4 (1999): Inside front cover.

2. Edward Blyth, 'Proceedings of the Asiatic Society of Bengal, for March, 1848: Report of Curator Zoological Department', *Journal of the Asiatic Society of Bengal* XVII, no. XV [Part I New Series] (1848): 247–55.

3. Bharat Bhushan, 'Rediscovery of the Jerdon's or Double-banded Courser Cursorius bitorquatus', *Journal of the Bombay Natural History Society* 83, no. 1 (1986): 1–14.

4. WhatsApp message from P. Jeganathan, who studied the ecology of (mainly habitat, distribution) the Jerdon's Courser for his PhD in the Sri Lankamakeswara Wildlife Sanctuary, Kadapa, from 2000 to 2008.

5. This is the Telugu name for the bird. Kalivi translates to 'carissa', the common bush of the region, under which the courser rests. Kodi translates to 'fowl'. So, the fowl that is found under kalivi bushes.

6.   William Wordsworth, 'Lines Composed a Few Miles above Tintern Abbey, On Revisiting the Banks of the Wye during a Tour, July 13, 1798', in *Lyrical Ballads* (London: J. & A. Arch, 1798).

7.   SoIB. 'State of India's Birds, 2020: Range, Trends and Conservation Status', India: The SoIB Partnership, 2020. 1–50.

8.   Although this phrase is widely ascribed to T.H. Huxley, this might as well be a mondegreen for his statement, 'birds are greatly modified reptiles' in *On the Classification of Birds; And on the Taxonomic Value of the Modifications of Certain of the Cranial Bones Observable in that Class, From the Proceedings of the Scientific Meetings of the Zoological Society of London, 1867* (London: The Zoological Society of London, 1867), 415–72.

9.   Sálim Ali, *The Book of Indian Birds*. 10th ed. (Bombay: Bombay Natural History Society, 1977).

10. David Lack, *Swifts in a Tower* (London: Methuen & Co., 1956).

11. Ali, *The Book of Indian Birds*.

12. Edward Hamilton Aitken [EHA], *The Common Birds of Bombay* (Bombay: Thacker & Co., 1900).

13. Ali, *The Book of Indian Birds*.

14. 'Grebes (Podicipedidae), version 1.0.', Birds of the World – Cornell Lab of Ornithology (website), last modified 4 March 2020, https://doi.org/10.2173/bow.podici1.01.

15. Henry David Thoreau, 'Walking', *The Atlantic Monthly*, June 1862.

16. Aitken, *The Common Birds of Bombay*.

17. Mark Cocker and David Tipling, *Birds and People*, 1st ed. (London: Jonathan Cape, 2013).

18. 'Pigeons – Everything There Is to Know about the Pigeon', Pigeon Control Resource Centre 2009 (website), https://www.pigeoncontrolresourcecentre.org/html/about-pigeons.html.

19. The reader will find several lists of personalities through a simple Google search.

20. Interstitial lung disease (ILD affects the interstitium, a lace-like network of tissue, which is a part of the lungs' anatomical structure. This gets scarred, causing what is called idiopathic [=the cause of which is unknown] pulmonary fibrosis [=scarring]).

21. Shirish Vaktania, 'Mumbai: Police Book Two for Feeding Pigeons, One for Selling Bird Food'. *Mid-day*, 23 January 2021. https://www.mid-day.com/mumbai/mumbai-news/article/mumbai-police-book-two-for-feeding-pigeons-one-for-selling-bird-food-23156762.

22. PTI, 'Can't Feed Birds from Balcony of Flat & Create Nuisance for Others: SC', *The Times of India*, 19 March 2019. https://timesofindia.indiatimes.com/city/mumbai/cant-feed-birds-from-balcony-of-flat-create-nuisance-for-others-sc/articleshow/68474482.cms.

Birding

23. 'Asian Waterbird Census', Wetlands International (website). https://south-asia.wetlands.org/our-approach/healthy-wetland-nature/asian-waterbird-census/.

24. William Wordsworth, 'To the Skylark', in *Poetical Works* (London: Longman, 1827)

25. Percy Bysshe Shelley, 'To the Skylark', in *Prometheus Unbound: A Lyrical Drama in Four Acts with Other Poems* (London: C and J Ollier, 1820).

26. The trip took place in November 1998.

27. WWF-India and A.P. State Office, 'Sacred and Protected Groves of Andhra Pradesh', 1996. https://archive.org/stream/sacredgrovesap/sacred%20groves%20ap_djvu.txt.

28. 'And the narrowest hinge in my hand puts to scorn all machinery, / And the cow crunching with depress'd head surpasses any statue, / And a mouse is miracle enough to stagger sextillions of infidels', Walt Whitman, 'Song of Myself', in *Leaves of Grass: The First (1855) Edition*, stanza 31 (Reprint, New York: Penguin Classics, 1986). Citations refer to the Penguin edition.

29. Whitman, 'Song of Myself', stanza 31.

30. E.M. Forster, *A Passage to India* (Reprint, London: Penguin Modern Classics, 1983). Citation refers to the Penguin edition.

31. Bo Beolens, Michael Watkins and Michael Grayson, *The Eponym Dictionary of Amphibians* (Exeter: Pelagic Publishing, 2013).

32. Henry Sullivan Thomas, *The Rod in India*, 3rd ed. (London: W. Thacker & Co., 1897). https://www.indianculture.gov.in/rarebooks/rod-india-0.

33. Source J.B.N.H.S. (1946): 46(2): 406–07. With a photograph!

34. William Wordsworth, 'The World is Too Much With Us,' in *Poems, In Two Volumes* (London: Longman, Hurst, Rees and Orme, 1807).

35. Walt Whitman, 'Birds–And a Caution', from *Specimen Days* in *Complete Prose Works* (Philadelphia: David McKay, 1892).

Birders

36. Isaac Walton, *The Compleat Angler* (London: Richard Marriot, 1653).

37. Sálim Ali, *The Fall of a Sparrow* (India: Oxford University Press, 1985).

38. Helen Macdonald, *H is for Hawk* (London: Jonathan Cape, 2014).

39. Harriet Ronkon Lynton and Mohini Rajan, *The Days of the Beloved* (Hyderabad: Orient Blackswan, 2012).

My Kind of Birding

40. The title is taken from Sharman Apt Russell's *Diary of a Citizen Scientist: Chasing Tiger Beetles and Other New Ways of Engaging the World* (Corvallis: Oregon State University Press, 2014).

41. Sahas Barve, T.R. Shankar Raman, Aparajita Datta and Girish Jathar, 'Guidelines for Conducting Research on the Nesting Biology of Indian Birds', *Indian BIRDS* 16, no.1 (13 July 2020): 10–11.

42. Aldo Leopold, 'The Geese Return', in *A Sand County Almanac: And Sketches Here and There* (Oxford: Oxford University Press, 1949).

Acknowledgements

I must thank T.R. Shankar Raman (Sridhar) first, for alerting me about the new imprint, Indian Pitta, that Anita Mani had started, and suggesting I send my essays to her for consideration. Several people have been responsible for helping me publish some of these essays in newspapers and magazines, and online, and I would like to record my sincere thanks to them: A.T. Jayanti, P. Balu, Chawla, Bittu Sahgal, Bikram Grewal, Sejal Mehta, and Radha Rangarajan. I would also like to thank long-time birding buddies Binod C. Choudhury, late Siraj Ahmed Taher, Rajeev Mathen Mathew, Mukund S. Kulkarni, N. Shiva Kumar, and Suhel Quader for joyous times, many of which were under the banner of the erstwhile Birdwatchers' Society of Andhra Pradesh. I have read and admired the amazing natural history writing of Mark Cocker and am hugely grateful to him for writing the Foreword. My editor, Anita Mani, has been an exceptional and understanding reader. I have benefited immensely from her inputs, support, and patience, and this book has improved manyfold due to her attention. I thank the Juggernaut team for all the work they have done behind the scenes to make this book possible. I thank Sangeetha Kadur for her delightful illustrations, which have illuminated the book. And finally, Smriti, my first reader, critic, cheerleader, and companion – you suffuse every page.

A Note on the Author

Aasheesh Pittie is the editor of the ornithological journal *Indian BIRDS* [www.indianbirds.in]. His interest in birds dates back four decades and has been the engine behind books such as *Birds in Books: Three Hundred Years of South Asian Ornithology* (Permanent Black, 2010) and *The Written Bird: Birds in Books 2* (New Ornis Foundation, 2022). Aasheesh has also compiled a searchable bibliographic database of over 35,000 works on South Asian ornithology [www.southasiaornith.in].

About Indian Pitta

Indian Pitta is India's first dedicated book imprint for bird lovers, conservationists, and policy makers. Our books about birds and natural history go beyond field/identification guides, to explore the bigger mosaic of habitats, ecosystems, and human interactions that touch the lives of birds. Successful conservation programmes, troubling environmental challenges, personal exploration of a landscape, deep dives into the ecology of a species, the quest for a rare species, and the sheer joy of birding – these are some of the ideas that you can expect to explore within the pages of our books.

Upcoming Titles in 2023

The Birds of the Delhi Area by Sudhir Vyas

Delhi is one of the most bird-rich national capitals in the world with more than 450 bird species. Sudhir Vyas' *The Birds of the Delhi Area* offers an updated, thoroughly researched, and annotated checklist of the birds found in the National Capital Territory and the surrounding areas that can be covered within a day's birding excursion from the city. This book is not a field guide. Instead, it will attempt to provide a better insight into broader aspects of the birds seen – beyond just identification – to contextualize their presence in the Delhi area, their habitat and distribution, and to sum up the current state of our knowledge of their status and changes in status over the past several decades. Richly informative and authoritative, *The Birds of the Delhi Area* is an indispensable book for every birdwatcher looking for birds in and around Delhi as well as those interested in the avi-fauna of one of India's bird-rich regions.

Women in the Wild edited by Anita Mani

How many Indian wildlife biologists can you name? How many Indian women wildlife biologists can you name?

Yet, there are several, and their lives and work have been extraordinary.

There's 'Turtle Girl', J. Vijaya, one of India's first female herpetologists whose research into the killing of Olive Ridley turtles led to Prime Minister Indira Gandhi banning the turtle trade.

There's Divya Mudappa, a biologist helping rewild fragments of the shola grasslands of the Western Ghats.

There's Uma Ramakrishnan, whose work is helping us decipher tiger populations on a scientific basis.

And many more such extraordinary women with equally extraordinary stories.

Written by natural historians, including Prerna Bindra, Neha Sinha, Ananda Banerji, and Zai Whitaker, _Women in the Wild_ will not only explore the details of their work – and their impact – but also what it took for them to get there.

Inspiring and eye-opening, _Women in the Wild_ will showcase the untold stories of some of India's most brilliant women wildlife scientists.

Birder on the Road by Shashank Dalvi

Wildlife biologist Shashank Dalvi's _Birder on the Road_ recounts his year-long trip across India to watch and record 1,128 species of birds. Travelling the length of India by plane, train, car, bus, boat, bike, and on foot, he describes the people, places, his mishaps, and lucky breaks with the same attentive, charming eye he uses to talk about his beloved birds. As much about India as its extraordinarily rich and diverse bird life, _Birder on the Road_ will be a travel and wildlife classic.